Determination of sequences in RNA

LABORATORY TECHNIQUES IN BIOCHEMISTRY AND MOLECULAR BIOLOGY

Edited by

T. S. WORK – *N.I.M.R.*, *Mill Hill, London*
E. WORK – *Imperial College, London*

Advisory board

G. POPJAK – *U.C.L.A.*
S. BERGSTROM – *Stockholm*
K. BLOCH – *Harvard University*
P. SIEKEVITZ – *Rockefeller University*
E. SMITH – *U.C.L.A.*
E.C. SLATER – *Amsterdam*

NORTH-HOLLAND PUBLISHING COMPANY – AMSTERDAM · LONDON
AMERICAN ELSEVIER PUBLISHING CO., INC. – NEW YORK

DETERMINATION OF SEQUENCES IN RNA

G. G. Brownlee

Medical Research Council,
Laboratory of Molecular Biology,
Hills Road, Cambridge, England

1972
NORTH-HOLLAND PUBLISHING COMPANY – AMSTERDAM · LONDON
AMERICAN ELSEVIER PUBLISHING CO., INC. – NEW YORK

© *1972 North-Holland Publishing Company*

All rights reserved. No part of this publication may be reproduced, stored in a retrieval system, or transmitted, in any form or by any means, electronic, mechanical, photocopying, recording or otherwise, without the prior permission of the copyright owner.

Library of Congress Catalog Card Number: 76-157037
ISBN North-Holland – series 0 7204 4200 1
  – part 3.I: 0 7204 4209 5
ISBN American Elsevier: 0 444 10102 0

63 illustrations & graphs, 30 tables

Published by:
NORTH-HOLLAND PUBLISHING COMPANY – AMSTERDAM

Sole distributors for the U.S.A. and Canada:
AMERICAN ELSEVIER PUBLISHING COMPANY, INC.
52 VANDERBILT AVENUE, NEW YORK, N.Y. 10017

This book is the pocket-edition of Volume 3, Part I, of the series 'Laboratory Techniques in Biochemistry and Molecular Biology'.

Volume 3 of the series contains the following parts:

Part I Determination of sequences in RNA, by G.G. Brownlee

Part II Techniques of lipidology: isolation, analysis and identification of lipids, by M. Kates

---

Printed in The Netherlands

Contents

Preface 8

List of abbreviations 9

Chapter 1. Introduction 11

1.1. Historical aspects 11
1.2. General approach to sequence determination 14
1.3. Nomenclature. 16

Chapter 2. Sequence of non-radioactive RNA 18

2.1. Introduction 18
2.2. Enzymic digestion 21
  2.2.1. Complete digestion by T_1-ribonuclease 21
  2.2.2. Complete digestion by T_1-ribonuclease and alkaline phosphatase . . . 22
  2.2.3. Complete digestion by pancreatic ribonuclease 22
  2.2.4. Partial digestion 23
    2.2.4.1. Partial T_1-ribonuclease 23
    2.2.4.2. Partial pancreatic ribonuclease 24
2.3. Fractionation of oligonucleotides 25
  2.3.1. Complete T_1-ribonuclease, and combined T_1-ribonuclease and alkaline
    phosphatase digests. 25
  2.3.2. Complete pancreatic ribonuclease digests 28
  2.3.3. Partial enzyme digests 28
    2.3.3.1. 'Half' molecules produced by partial T_1-RNase 28
    2.3.3.2. Extensive partial T_1-RNase digests 30

2.3.3.3. Partial P-RNase digests	34
2.4. Desalting of nucleotide peaks from columns	35
2.4.1. Gel filtration	35
2.4.2. Adsorption onto DEAE-cellulose	36
2.4.3. Dialysis	37
2.5. Yields of isolated oligonucleotides	37
2.6. Sequence in oligonucleotides	38
2.6.1. Sequence methods	42
2.6.1.1. Alkaline hydrolysis	43
2.6.1.2. T_2-RNase	46
2.6.1.3. Snake venom phosphodiesterase	48
2.6.1.4. Pancreatic RNase digestion of T_1-oligonucleotides and T_1-RNase digestion of P-RNase oligonucleotides	48
2.6.1.5. U_2-RNase digestion of T_1-end products	49
2.6.1.6. Partial digestion with snake venom phosphodiesterase	49
2.6.1.7. Micrococcal nuclease	51
2.6.1.8. Polynucleotide phosphorylase	51
2.6.1.9. Spectral identification	52

Chapter 3. High-voltage paper electrophoresis *54*

3.1. Introduction	54
3.2. High-voltage electrophoresis	54
3.3. The 'hanging' electrophoresis tank	55
3.4. 'Up-and-over' tanks	56
3.5. Power supply and safety precautions	58
3.6. Loading, buffers and operating	64

Chapter 4. A two-dimensional ionophoretic fractionation method for labelled oligonucleotides. *67*

4.1. Introduction	67
4.2. Enzymatic digestion	69
4.3. Two-dimensional ionophoretic fractionation procedure	70
4.4. Structure of nucleotides	75
4.4.1. Composition	76
4.4.2. Digestion with enzymes	77
4.4.2.1. Pancreatic ribonuclease	77
4.4.2.2. T_1-RNase	78
4.4.2.3. Micrococcal nuclease	78
4.4.2.4. Venom phosphodiesterase	80
4.4.2.5. Partial digestion with spleen phosphodiesterase	81

4.5. Application of the two-dimensional method for fingerprinting 16S and 23S rRNA	82
4.5.1. T_1-RNase digests	83
4.5.2. Pancreatic RNase digests	89
4.6. Results of partial spleen phosphodiesterase digestion of oligonucleotides	89
4.6.1. Partial spleen phosphodiesterase digestion of oligonucleotides derived from complete T_1-RNase digestion	90
4.6.2. Nucleotides from pancreatic ribonuclease digests	96
4.7. Conclusion	99

Chapter 5. Sequence of 5S RNA 100

5.1. Introduction	100
5.2. Complete T_1-RNase digestion	103
5.2.1. Secondary splits	107
5.2.2. Partial digestion of large oligonucleotides with snake venom phosphodiesterase	109
5.2.3. End groups	117
5.3. Complete pancreatic RNase digestion	118
5.3.1. Partial digestion with spleen phosphodiesterase	120
5.4. Relative molar yields and discussion	122
5.5. 'Homochromatography' and fractionation of partial enzymatic digests	130
5.5.1. A two-dimensional procedure using DEAE-paper	132
5.5.2. A two-dimensional procedure using DEAE-cellulose thin layers	136
5.5.2.1. Preparation of thin layers	136
5.5.2.2. Fractionation procedure	137
5.5.2.3. Homomixtures	140
5.5.2.4. Elution and analysis	141
5.6. Partial enzymatic digestion	142
5.6.1. Extensive digestion with T_1-RNase	143
5.6.2. 'Half-molecules' with T_1-RNase	148
5.6.3. Pancreatic RNase	150
5.6.4. Spleen acid RNase	155
5.7. Chemical blocking with water-soluble carbodiimide	155
5.8. Chemical blocking by partial methylation	165
5.9. Derivation of the sequence by overlapping	173
5.9.1. 5'-terminal sequence (residues 1-16)	175
5.9.2. 3'-terminal sequence (residues 103-120)	176
5.9.3. Residues 76-105	176
5.9.4. Residues 14-67	177
5.9.5. Residues 65-79	179
5.10. Discussion	180

5.10.1. Sequence variations	180
5.10.2. Secondary structure	181
5.10.3. Sequence homologies	185

Chapter 6. Methods for sequencing large oligonucleotides isolated as end-products of T_1-RNase digestion 187

6.1. Introduction	187
6.2. Isolation of large oligonucleotides	189
6.3. Deduction of the sequence of the oligonucleotide AAUUAACUAUUCCAAUUUUCG	192
6.4. Deduction of the sequence of the oligonucleotide AUAUUUCAUACCACAAG	194
6.5. Cleavage next to C's	196
6.6. Cleavage next to A's with RNase U_2	199
6.7. Partial spleen acid RNase digestion	200
6.8. Partial pancreatic RNase digestion	201

Chapter 7. Minor bases and tRNA 202

7.1. Introduction	202
7.2. Purification of ^{32}P-tRNA	203
7.3. Minor bases	204
7.4. Identification of minor bases in tyrosine suppressor tRNA of *E. coli*	208
7.5. Complete hydrolysis with T_2-RNase	210
7.6. A cross-linkage in *E. coli* tRNA$_1^{\text{Val}}$	211

Chapter 8. Limited sequence objectives: end-groups of RNA and sequences adjacent to minor bases in ribosomal RNA 213

8.1. Introduction	213
8.2. End groups in RNA	214
8.2.1. 5'-end nucleotide	214
8.2.2. 5'-end oligonucleotide	216
8.3. 3'-terminal oligonucleotide – a 'diagonal' method	218
8.4. Sequence around methylated bases in 16S and 23S RNA of *E. coli*	220

Chapter 9. In vitro labelled ^{32}P-RNA 225

9.1. Introduction	225
9.2. Polynucleotide phosphokinase labelling	227
9.3. Synchronised synthesis of RNA using purified replicase	231

Acknowledgements *235*

Appendices. *237*

1 Tentative rules for representation of nucleosides and polynucleotides 237
2 Addresses of suppliers of enzymes, electrophoresis supports and fine chemicals, etc. 239
3 Addresses of suppliers of equipment 240
4 ε_{mM} for some nucleotides and nucleosides at pH 7.0 and 260 mμ 241
5 Preparation of radioactively labelled RNA 241
6 Sequence of some oligonucleotides from rRNA 250
7 Summary of conditions of enzyme digestion and of fractionation systems for radioactive sequencing 250

References *256*

Subject index *261*

Dedicated to my mother

Preface

The aim of this monograph is to survey the methods that have recently been developed for the determination of the sequence of small RNA molecules. I have attempted to give the necessary experimental details for a reader to reproduce these methods in the laboratory and apply them to his own particular sequence problem. The chapter on electrophoresis equipment and technique will, I hope, be especially useful to those envisaging setting up the equipment for sequence determination. In this book, I have covered both the classical methods and the newer radioactive methods. However, my emphasis is naturally towards the latter as it is with these methods that I am most familiar. The recent spate of reports in scientific journals of the sequence of various transfer RNA molecules by radioactive methods have shown that this is a powerful and rapid approach. However, here I have not concerned myself with discussing the results of sequence work and they are only discussed in so far as is necessary to illustrate the methodology of sequence determination. Nor is any reference made to the methods of sequence determination of DNA, in which there has been less progress than with RNA. At the time of writing the longest known RNA sequence is that of the 6S RNA determined by the radioactive approach. It is certain, however, that larger sequences will be established by using these methods. Perhaps we may conjecture that the time will not be too distant when we may know the entire RNA sequence of the ribosome.

List of abbreviations
(excluding those defined in appendix 1 and table 7.2, p. 205)

tRNA	Transfer RNA.
rRNA	Ribosomal RNA.
mRNA	Messenger RNA.
G*	Unknown derivative of guanosine.
Py	Any pyrimidine nucleoside.
Pu	Any purine nucleoside.
R_u	Chromatographic or electrophoretic mobility of an unknown compound relative to uridine $2'(3')$-phosphate $=1.00$.
P_i	Orthophosphate.
P-RNase (RNase A)	Pancreatic ribonuclease.
T_1-RNase (RNase T_1)	Takadiastase T_1 ribonuclease.
T_2-RNase (RNase T_2)	Takadiastase T_2 ribonuclease.
U_2-RNase (RNase U_2)	Ustilago U_2 ribonuclease.
DNase	Pancreatic deoxyribonuclease.
mA	Milli-ampères.
C	Curie.
dpm	Radioactive disintegrations per minute.
p.f.u.	Plaque-forming units.
u	Units.
EDTA	Ethylenediaminetetraacetic acid.
Tris	Tris-(hydroxymethyl)-aminomethane.

SDS	Sodium dodecyl sulphate.
SSC	0.15 M sodium chloride–0.015 M sodium citrate, pH 7.0.
BBOT	2,5-bis-(2(5-tert-butylbenzoxazolyl))-thiophene.
CMCT	N-cyclohexyl-N'-(β-morpholinyl-(4)-ethyl)-carbodiimide-methyl-p-toluene sulphonate.
TEC	Triethylamine bicarbonate.
BSA	Bovine serum albumin.
M.W.	Molecular weight.
R_f	Chromatographic mobility expressed relative to rate of front.
rpm	Revolutions per minute.

1.1. Historical aspects

Although nucleic acids and the four common bases (fig. 1.1) derived from them have been known to chemists since the last century, there was no precise knowledge of their function until 1944 when Avery et al. showed that DNA was the essential component of the transfor-

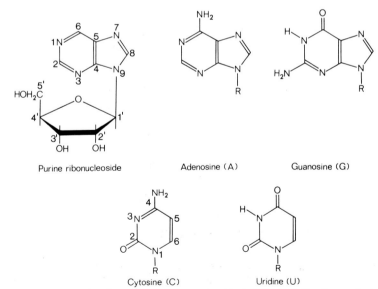

Fig. 1.1. Formula of the four common mononucleosides in RNA, including the numbering of the carbon atoms.

ming factor in pneumococci, and thus carried inheritable information. It required further careful chemical analysis, coupled with advances in paper and ion-exchange chromatography, to prove that the monomeric units of nucleic acids were joined together by a 3'-5' phosphodiester linkage (fig. 1.2) between adjacent ribose moieties (e.g. see

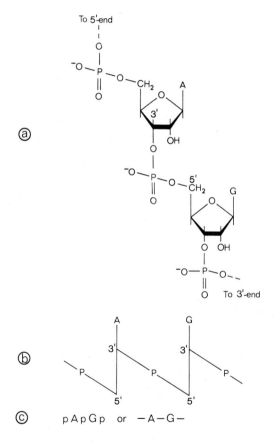

Fig. 1.2. Linkage and nomenclature in RNA. (a) shows the ribose-phosphate backbone which may also be drawn as in (b) or omitted altogether as in (c). Notice that in all cases the phosphodiester bridge is linked to a 3'-C-atom on its left and a 5'-C-atom on its right.

Brown and Todd 1952). However, it was not clear until 1953, when Watson and Crick correctly deduced the fibre structure of DNA, that a precise *sequence* of bases was important in maintaining the double-helical structure of the DNA. In particular, the structure was significantly stabilised by hydrogen bonds between specific complementary bases: adenine with thymine, and guanine with cytosine (fig. 1.3).

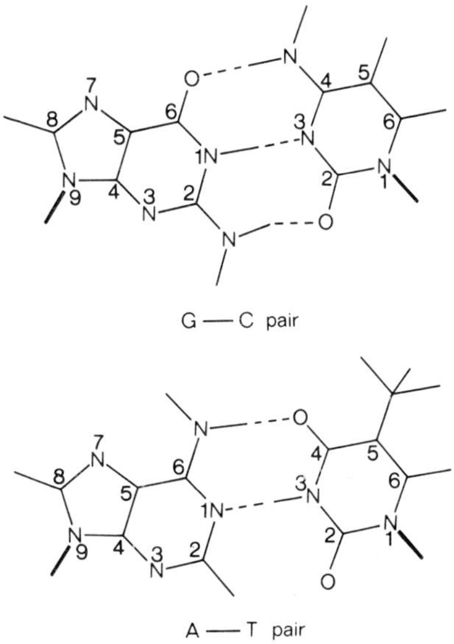

Fig. 1.3. Base pairs in DNA drawn approximately to scale. The hydrogen bonds are dotted and the glycosidic linkages are thickened. A–T is replaced by A–U in RNA. G–U ('wobble' pair) occurs in RNA and is rather similar to the G–C pair shown except that the third (uppermost) hydrogen bond is not made.

This specific pairing of bases in the double helix suggested to Watson and Crick (1953) a model for replication involving these base pairs by which the sequence was accurately conserved. Only later did Kornberg and his collaborators (Lehman et al. 1958) isolate a DNA polymerase from *E. coli* which had the correct properties for accurately copying DNA. The implications of these observations were quite clearly that a nucleotide sequence defined, in an as yet unknown way, the complete phenotype. RNA (which is not usually the gene) must also have a defined sequence as it, also, is synthesised by a copying of (usually) DNA so as to form a complementary sequence to one of the DNA strands in the presence of an RNA polymerase (Hurwitz et al. 1961). Indeed, from our knowledge of the defined amino-acid sequence of any given protein, and a knowledge that it is synthesised on a mRNA template (Volkin and Astrachan 1956) it would be surprising if anything less than a precise sequence existed for mRNA. It is abundantly clear that nucleic acids are exposed to the process of natural selection and that the sequences which exist are those most adapted to function. Thus, in order to define and understand their function we must study their sequence.

1.2. General approach to sequence determination

Progress in the determination of nucleotide sequences in nucleic acids has lagged behind progress made in the sequence analysis of proteins. The basic approach is very similar for both polymers and requires the partial degradation of the molecule into smaller fragments of various sizes. After purification, the sequence of these smaller fragments is established by further degradative (usually enzymatic) methods. By analysing a sufficient number of fragments of various sizes, one hopes to accumulate sufficient information to logically deduce the entire sequence. The reason for the slow progress with nucleic acids has been partly due to experimental difficulties in obtaining homogeneous and undegraded low molecular-weight preparations of a given nucleic acid, and partly due to the lack of techniques for fractionation of the extremely complex mixture of related products in a partial digest. In addition, as there are only 4 basic units in nucleic acids compared to twenty amino acids in proteins, it is necessary to isolate, in a pure form, rather longer sequences for them to be unique. As with proteins, only a few of the known nucleic acids are small enough to be susceptible to the present methods of sequence analysis. Unfortunately the smallest known RNA molecule, from satellite tobacco necrosis virus, has over

1100 nucleotides per molecule (Markham 1963) and is not a good candidate for sequence work. However, it has been known for some time that the most promising nucleic acids were the transfer RNA molecules (about 80 residues long) and it is with these that most progress has been made.

As I outline more fully in ch. 2, the particular amino-acyl-specific tRNA in question must first be purified and only then may its sequence be determined. Holley et al. (1965a) in their classic work did this and successfully deduced the sequence of yeast alanine tRNA after separating fragments of the molecule by chromatography on DEAE-cellulose columns. In all this work, the chromatographic fractionation procedure on *columns* was undoubtedly the rate-limiting step and presented the greatest difficulty in sequence determination. However, by making use of uniformly ^{32}P-labelled nucleic acids and using two-dimensional fractionation methods on *modified paper*, with its high resolution, Sanger and his collaborators (Sanger et al. 1965; Sanger and Brownlee 1967) have to a large extent simplified these fractionation problems. Moreover, by using labelled nucleic acids many of the sequence procedures are also considerably simplified because of the sensitivity of the radioactive techniques. This radioactive approach turned out to be a break-through in methodology and allowed the rapid elucidation of sequences of a number of tRNA and 5S ribosomal RNA molecules. The two contrasting approaches are discussed in detail in the rest of this monograph, ch. 2 being devoted to a description of the classical methods of Holley and others, whilst the succeeding chapters describe the radioactive methods of Sanger and his colleagues. The evidence for one sequence, that of 5S RNA of *E. coli*, is presented in detail in ch. 5 so that the reader may be aware of the nature of the evidence required to deduce the structure of a molecule of this length (120 residues). The next chapters describe specialised problems of radioactive sequencing. Ch. 6 is devoted to a discussion of the sequence determination of very long end-products of T_1-ribonuclease digestion. In ch. 7 I discuss some of the difficulties and peculiarities of sequencing tRNA at the radioactive level. Ch. 8 describes two interesting applications of the radioactive methods which are not concerned with de-

Subject index p. 261

fining a total sequence. These are end-group determination and sequences containing methylated bases in high molecular weight ribosomal RNA. The last chapter describes some very recent applications of the radioactive sequence methods for sequencing RNA labelled in vitro. The results of sequence work are not discussed in this monograph.* A useful summary of radioactive sequence methods appears in appendix 7.

1.3. Nomenclature

One of the necessary facts of chemistry is the need to abbreviate the long formal names of compounds in order to simplify their representation. Although this book is concerned with practical techniques rather than nomenclature, the results of applying the techniques will be enumeration of nucleotide sequences. Representation is, therefore, of importance, and the reader will perhaps excuse a digression into a discussion of this problem.

The accepted way of presenting a sequence is a linear order of single capital letters C, A, G and U representing the four common nucleosides cytidine, adenosine, guanosine and uridine. These are joined together by hyphens representing the 3'-5' phosphodiester linkage if the sequence is known. Where the composition of an oligonucleotide is known, but the residue order is not, this is indicated by enclosing the residues of unknown order within brackets with a comma rather than a hyphen between each residue symbol. Minor nucleosides are represented as derivatives of the nucleosides above, e.g. N^7-methyl guanosine is represented as m^7G. Full rules for representation of minor nucleosides appear in the IUPAC-IUB tentative rules (J. Biol. Chem. (1966) *241*, 527) which have been adopted in this book and examples appear in table 7.2. An extract of these rules appears in appendix 1.

Sometimes, however, it is convenient to abbreviate the accepted nomenclature even further than is suggested in these rules. In sequence

---

* For a collection of some results of sequence determination see Sober (1968) and Zachau (1969).

work it is often convenient to use the accepted symbols for *nucleosides* as though they applied to *nucleotides*. This is especially true when sequencing ^{32}P-labelled nucleic acids where nucleosides are never observed. Thus a sequence written under IUPAC rules as A-C-A-C-G (or ApCpApCpGp) simply becomes ACACG. This abbreviation of the accepted nomenclature suffers from the disadvantage that it does not distinguish A-C-A-C-G from A-C-A-C-G-, for example, and the point must be clarified if this abbreviation is used. However, if we accept that abbreviation (as proposed in the IUPAC rules) was designed for clarity of representation and of reading, then it is perhaps justified to use a further abbreviation where this improves clarity. This is true of tables of partial sequences and especially of figures showing a 'fingerprint' of a nucleic acid with the sequence written beside the spots. Here, without doubt, hyphens complicate the representation unnecessarily. Moreover, in drawing out and comparing long sequences such as that of the various tRNA molecules, it is often very inconvenient to include hyphens which do not add any information to the representation and are often, therefore, omitted by authors. In this monograph I have not used hyphens except where it is necessary to distinguish oligonucleotides with or without a 3'-terminal phosphomonoester group, or where it is necessary for an understanding of the pattern of degradation of the oligonucleotide in question.

Subject index p. 261

Sequence of non-radioactive RNA

2.1. Introduction

The classical approach to sequence determination uses the ultraviolet absorption property of the aromatic ring on the base (maximum at 260 mμ) to detect the presence of the nucleoside or nucleotide. Because the sensitivity of detection is limited to the amount which can be detected by adsorption at 260 mμ, it is necessary to purify as much as 100 to 200 mg of a particular transfer RNA in order to be in a position to start serious sequence analysis. Moreover, the fractionation and purification on this scale is a long and laborious task and must usually involve: (1) A large-scale extraction of crude low molecular-weight RNA, arranged as a separate (even commercial) project. (2) A multi-step purification of the RNA of interest which uses a fractionation technique capable of taking up to 10 g of RNA. In general, countercurrent distribution has been used for this step, and repeated redistribution on a smaller scale can be used to obtain pure material, although a large number of transfers may be required. A discussion of the techniques used in purification of low molecular-weight RNA can be found in the monograph in this series by Matthews and Gould (in prep.) and therefore no further details are given here.

The sequence determination of non-radioactive RNA is also laborious, principally because many column runs are required both in the initial fractionation and in the subsequent analyses of oligonucleotides. The fractionation of some larger oligonucleotides (10 to 20 residues in length) may involve up to 3 sequential column-chromatographic steps before sufficient purity is reached for an unambiguous analysis. If there is sufficient yield of material at this stage, the analysis of such an oligonucleotide may then involve 2 further column runs. However, as discussed in § 2.6.1 the smaller oligonucleotides may be analysed by paper or thin-layer methods with advantage.

It is with these points in mind that the radioactive approach, with its much greater sensitivity, should be considered. Both the purification and sequence analysis can then be done by one person within a reasonable period of time. The purification is in some respects simpler because it is not necessary to process exceedingly large weights of material so that smaller scale methods can be used. In practice this means that column chromatography and even 'analytical' polyacrylamide gel electrophoresis with its high resolution may be used instead of countercurrent distribution. Sequence determination is simpler because paper methods may be used both in the fractionation of the initial nucleic acid digest and in its subsequent analysis.

Nevertheless a number of nucleotide sequences, specifically those of alanine, serine, tyrosine, valine and phenylalanine tRNA molecules of yeast and others now in progress, have been established by the non-radioactive approach. There are instances, also, where it may not be possible, or convenient, to obtain radioactive RNA, e.g. in the study of mammalian tRNA and in these cases the approach described in this chapter is applicable. However, it is probable that radioactive end-group labelling methods, which are discussed further in ch. 9, will make it possible to sequence non-radioactive nucleic acids by radioactive methods. Perhaps the only future requirement for large amounts of unlabelled RNA will be for physico-chemical studies and for crystallisation. In any case, the large amounts of RNA required would make it extremely difficult to obtain sequences of ribonucleic acids of chain length much greater than 100. In this chapter, therefore, discussion is limited to the methods developed for sequencing tRNA molecules.

The basic principles of sequence determination are, of course, the same whether the RNA is radioactive or not. The only difference is in the fractionation procedures employed. The rationale in sequence analysis is to degrade the molecule specifically using a base-specific

Subject index p. 261

ribonuclease, to separate the resulting small fragments from one another and to sequence them and determine their yields. These smaller fragments are then ordered into a unique total sequence by isolating and analysing progressively larger oligonucleotides obtained by partial digestion of the RNA using limiting amounts of enzyme. T_1-ribonuclease is the only guanine-specific ribonuclease that has been used extensively. (Other enzymes with this same specificity are now known, but have no advantage over T_1-RNase.) Pancreatic ribonuclease is specific for pyrimidine residues and may be used for obtaining sequences around clusters of purine residues and for cross-checking the results of the T_1-ribonuclease digest. Thus the usual procedure is to digest to completion with T_1-ribonuclease using conditions which give guanosine 3'-phosphate and oligonucleotides ending in guanosine 3'-phosphate, then digest to completion with pancreatic ribonuclease. After completion of the sequence analysis of the products and the measurement of yields the investigation is extended by partial digestion with either T_1- or pancreatic ribonuclease, and products are sought which enable one to construct a unique RNA sequence. The reader will notice that the terms 'complete' and 'partial' digestion with an enzyme such as T_1-RNase have a precise meaning. A complete digest gives end-products with only *one* guanine residue per oligonucleotide. A partial digest gives partial fragments having more than one guanine residue.

The main problem in sequence work is the fractionation of the enzymatic digestion products so that they are obtained pure for analysis. Where use has not been made of ^{32}P, ion-exchange chromatography on columns of DEAE-cellulose or on DEAE-Sephadex has been used in successful studies. Columns are normally run in the presence of 7 M urea following the observation by Tomlinson and Tener (1963) that secondary binding forces between the oligonucleotide and the exchanger are largely eliminated so that fractionation is primarily on the basis of net charge. To a close approximation, at *p*H values between 5.5 and 8.0, nucleotides will therefore be separated according to the number of phosphate groups or their chain length. However, 7 M urea does not completely eliminate all secondary binding forces particularly with

DEAE-cellulose, and this residual binding force may be exploited in the fractionation of enzyme digests so that when long columns (200 cm) are run, separation between oligonucleotides of equal chain length may be obtained. Oligomers which have been separated primarily according to size at neutral pH may be sub-fractionated by rechromatography on a second type of column, either on DEAE-Sephadex or on DEAE-cellulose at acid pH in the presence of 7 M urea (Rushizky et al. 1964). At pH 3.0, the amino group of cytosine ($pK_a \sim 4.3$) and of adenine ($pK_a \sim 3.7$) are both almost fully protonated whilst guanine ($pK_a \sim 2.3$) is almost completely uncharged. Thus the net charge, and therefore the separation, will be strongly influenced by the base composition of the oligonucleotide. A third type of column is the hot DEAE-cellulose column in 7 M urea (Apgar et al. 1966), which produces a higher resolution, mainly according to chain length, of the larger oligonucleotides.

The three types of column – neutral, acid and hot – are the basic fractionation methods available. Whether DEAE-Sephadex or DEAE-cellulose is used, the details of the length of column and other variables, depend on the particular separation problem. The following practical details give the most common procedures. Most are taken direct from the experimental sections of research papers without modification. Some earlier fractionations were done on columns without 7 M urea, but recovery of the larger fragments was poor.

2.2. Enzymic digestion

2.2.1. Complete digestion by T_1-ribonuclease

T_1-RNase (7000 units $\sim$ 1 mg, see Appendix 2) is used without further purification. Conditions should be such that all oligonucleotides end in guanosine-3'-phosphate, the only 2',3' cyclic phosphates present being G$\rangle$ (and pG$\rangle$). For overnight digestion the following conditions of Zamir et al. (1965) may be used: 10-40 mg of RNA were digested in 1 ml of 0.02 M sodium phosphate buffer, pH 7.5, using approximately 15 units of enzyme per mg of RNA. A drop of chloroform was added to prevent growth of micro-organisms and incubation was carried out for 24 hr at 37 °C. Shorter digestion times may be convenient and better as there is less chance of 'over-digestion'. For this, proportionately larger amounts of enzyme are needed (see enzyme digestion of radioactive RNA as in § 4.2). After incubation, the sample is diluted with 10 times its volume of 7 M urea and loaded immediately onto a column for fractionation.

2.2.2. *Complete digestion by T_1-ribonuclease and alkaline phosphatase*

This type of digestion is designed to give oligonucleotides lacking a 3'-terminal phosphate group. This is an important type of degradation because the products are then substrates for partial digestion with snake venom phosphodiesterase (see § 2.6.1.6) which is the principal method used for sequencing the larger T_1-oligonucleotides. Electrophoretically purified bacterial alkaline phosphatase (appendix 2) may be used without further purification. The following conditions of Zachau et al. (1966) are suitable: 640 A_{260}-units* of serine tRNA I + II were digested for 2 hr at 45 °C with 0.4 A_{280}-units (approx. 0.4 mg) of T_1-RNase in 0.01 M Tris-chloride, pH 7.5, final volume 5 ml. The solution was made pH 8.0 by adding 0.5 ml of 1 M Tris-chloride, 0.1 M $MgCl_2$ (pH 8.0) and 0.4 A_{280}-units of bacterial alkaline phosphatase were added. Incubation was continued for a further 2 hr at 37 °C. The solution may then be diluted to 10 times with the starting buffer containing 7 M urea for fractionation on a neutral type DEAE-cellulose column. Zachau et al. (1966) chose to use a system lacking urea. Alternative conditions for digestion are given in § 4.2.

2.2.3. *Complete digestion by pancreatic ribonuclease*

Again conditions should be such that all oligonucleotides terminate in pyrimidine 3'-phosphates, the only cyclic phosphates being the mononucleoside 2', 3'-cyclic phosphates. Many workers use unne-

---

* One absorbancy unit (A) is that weight of material which gives an extinction of 1.0 at the stated wavelength (in mμ) in a cell of 1 cm path length, when dissolved in 1.0 ml of water. For RNA, 1 A_{260}-unit = 40 μg; for protein, 1 A_{280}-unit = 1 mg very approximately.

cessarily violent conditions with the result that the larger oligonucleotides are present in low yield because pancreatic ribonuclease is known to split slowly between purine residues. Normally the digestion is terminated by cooling, adding urea and immediately loading onto a prepared column. Pancreatic ribonuclease (appendix 2) may be used without further purification: if necessary a further purification step may be used involving heating at 100 °C at pH 4.5 and rechromatography on carboxymethyl-cellulose columns (Dutting et al. 1966). The following conditions (Zachau et al. 1966) are suitable: 268 A_{260}-units (approx. 25 mg) of serine tRNA were digested in 2 ml for 2 hr at 45 °C with 0.25 A_{280}-units (approx. 0.25 mg) pancreatic RNase in 0.01 M Tris-chloride at pH 7.7. (Enzyme to substrate ratio is approximately 1 to 100.) Alternatively the following more vigorous conditions of Holley et al. (1961) may be used: 7.7 mg of alanine tRNA was dissolved in 3.0 ml water and the solution mixed with 0.37 ml of 0.1 M sodium phosphate buffer (pH 7.0), 0.37 ml of 1 mg/ml solution of crystalline pancreatic RNase and 0.15 ml of chloroform. The mixture was incubated for either 4 hr or 14 hr at 37 °C. (Enzyme to substrate ratio is approximately 1 to 20.) The shorter time was necessary in order to obtain good yields of the larger oligonucleotides.

2.2.4. Partial digestion

Both T_1- and pancreatic ribonuclease have been used, although some workers have studied only the former. Partial T_1-RNase digestion is therefore discussed first.

2.2.4.1. Partial T_1-ribonuclease. The conditions depend on which range of partial products are required. For very limited digestion where 'half' molecules are required and where unchanged nucleic acid remains, the conditions of Penswick and Holley (1965) may be attempted. Notice that the digestion is carried out in the presence of Mg^{2+}. This uses a 4-min digestion at 0 °C in 1.1 ml of 0.02 M $MgCl_2$, 0.02 M Tris-chloride pH 7.5 and 25 units of T_1-RNase per 125 A_{260}-units (approx. 5 mg) of alanine tRNA. The reaction was terminated and the enzyme removed by mixing the solution with 2 ml of phenol

Subject index p. 261

(previously equilibrated with the Tris-magnesium buffer) at 0 °C and the lower phenol layer was removed by centrifugation. The extraction of the T_1-RNase was repeated twice with 2-ml volumes of phenol. The aqueous layer was then extracted five times with ether to remove traces of phenol and then freed of ether by bubbling nitrogen through the solution. The solution was made 7 M with respect to urea and then diluted 10 times with the starting buffer containing 7 M urea for application to a *neutral* or *hot* DEAE-cellulose column. A procedure similar to this also gave 'half' molecules from tyrosine tRNA (Madison et al. 1967a), and valine tRNA (Baev et al. 1967), but not from phenylalanine tRNA (RajBhandary et al. 1967), all from yeast.

For a more extensive partial digestion with T_1-RNase the following conditions (Apgar et al. 1966) were suitable: 30 mg of alanine tRNA were dissolved in 15 ml of 0.2 M Tris-chloride buffer, pH 7.5, cooled to 0 °C and combined with 2.7 ml (6750 units, approx. 1 mg) of a cold solution of T_1-RNase. After 1 hr at 0 °C the digestion was terminated and the enzyme removed by phenol as described in the preceding paragraph. 31.5 g of urea was added to the aqueous layer followed by water to a final volume of 75 ml (to make 7 M with respect to urea) before loading onto a neutral DEAE-cellulose column. Partial digestion under identical conditions except for the presence of 0.02 M Mg^{2+} has also been described (Dutting et al. 1966).

2.2.4.2. Partial pancreatic ribonuclease. Dutting et al. (1966) used a 30-min digestion at 0 °C with 1 part of enzyme to 200 parts of substrate in the presence of Mg^{2+} at pH 7.5. (More extensive degradation can be obtained by increasing the ratio of enzyme to substrate up to 1 : 40 (see Chang and RajBhandary 1968). 578 A_{260}-units (approx. 20 mg) of serine tRNA in 4 ml of 0.1 M Tris-acetate, pH 7.5, 0.01 M magnesium acetate was cooled to 0 °C and digested with 0.094 A_{280}-units (approx. 0.1 mg) of P-RNase. After 30 min at 0 °C the pH was adjusted to pH 6.0 with acetic acid and 0.3 ml of 'Macaloid' suspension (appendix 2) (7.5 mg/ml) in 0.01 M sodium acetate, pH 6.0 (prepared according to Stanley and Bock (1965)) was added and shaken for 2 min in a Vortex mixer. Pancreatic RNase was adsorbed to the Macaloid which was removed by centrifugation at 12 000 rpm for 5 min and the supernatant was retreated twice more with 0.3 and 0.1 ml Macaloid as above. The three Macaloid precipitates were combined and washed twice with 1 ml water and the washings and combined supernatants were adjusted to pH 7.8 with dilute ammonia. The solution was passed slowly through a column (0.8 × 8 cm) of Chelex-100 ammonium form (appendix 2). The Mg^{2+}-free filtrate (35 ml) was concentrated by rotary evaporation before application to a neutral type DEAE-cellulose column. Instead of the Chelex-100 step, the digest may be treated with 1 M NaCl to a final concentration of 0.1 M and then with 2 volumes of 95% ethanol; the oligonucleotide precipitate is collected by centrifugation, washed in 95% ethanol, recentrifuged and traces of ethanol dried off in a desiccator.

2.3. Fractionation of oligonucleotides

The following is a summary of some of the procedures used. The principles have been discussed in § 2.1.

2.3.1. Complete T_1-ribonuclease, and combined T_1-ribonuclease and alkaline phosphatase digests

A neutral type of DEAE-cellulose column run in the presence of 7 M urea is probably best. Long narrow columns (100-200 cm × 0.35-1.0 cm) are required in order to obtain maximum resolution, although not all workers use the longer and narrower columns which are difficult to prepare and pack (see Peterson, this series, vol. 2, part II, ch. 6).

Commercial DEAE-cellulose must be washed in 0.5 N HCl (or acetic acid) and 0.5 N NaOH (or 1 M NaCl) before use (Peterson and Sober 1962). The finer particles which tend to block up the column must also be removed. Schleicher and Schüll, 70, medium has been used by Holley and his associates. Whatman DE-52 is also suitable (Dr. Doctor, personal communication). DEAE-cellulose from Serva Entwicklungslabor, Heidelberg, 200-325 mesh, 80-130 DIN; and DEAE-cellulose (Cellex-D) from California Corp. Biochem. Res. have been used by Zachau and his associates. The following is a typical procedure taken from Apgar et al. (1966). DEAE-cellulose (Schleicher and Schüll, 70, medium, 0.9 meq/g with fines removed by suspension in water) was washed with 0.2 M acetic acid, water, the maximum concentration of the elution gradient and finally in water, and was stored in water, preferably for a month (stated to improve resolution). Long narrow columns are conveniently packed in two sections using approximately 4-ft sections of Pyrex tubing of the required internal diameter – 0.35 cm. Two such columns were supported vertically, funnels were attached to the tops of the tubes and short, tapered glass tips containing glass wool were attached to the bottom of the tubes with short pieces of Tygon tubing. The glass tubing was packed with DEAE-cellulose which had been centrifuged (or refiltered) and washed and resuspended with a freshly prepared solution of 7 M urea in 0.02 M Tris-chloride, pH 8.0. The absorbent was allowed to settle by gravity. Any clumps which formed in the tube above the column were broken up with a stainless steel or Nichrome wire in such a way that it was not pushed into that part of the column that had already been packed. The two 4-ft sections were joined in series using a short length of Tygon tube, and a clean funnel was attached to the top of the column before loading the sample. The sample was washed into the column (run at room temperature) with 2 ml of the starting buffer and then eluted with a linear gradient produced from 225 ml of 7 M urea, 0.02 M Tris-chloride, pH 8.0 and 225 ml of 0.3 M sodium chloride in 7 M urea, 0.02 Tris-chloride, pH 8.0. The flow rate was approximately 10 ml/hr (a pump may be used, or hydrostatic pressure) and approximately 1-ml fractions were collected. A typical profile is shown in fig. 2.1 (upper half) (Apgar et al. 1966). Most peaks are pure, but where heterogeneity is found further resolution may be obtained by high-voltage paper electrophoresis, as described in ch. 3, at pH 2.7 on Whatman No. 3 MM paper. Fractionation at other pH's may be useful, for example at pH 3.5 (5% acetic acid adjusted with ammonia to pH 3.5). In order to improve the resolution on the paper the 3'-terminal phosphate may be removed by digesting with bacterial alkaline phosphatase (§ 2.6.1.6) before running on the paper.

For the larger oligonucleotides (chain length 4 or more, or 5 or more if dephosphorylated), which tend to streak on paper, it may be necessary to rechromatograph some of the peaks on DEAE-Sephadex at an acid pH (Rushizky et al. 1964) as follows: DEAE-Sephadex A-25 was freed of fines by repeated decantation of material that did not settle in about 30 cm in water in 5 min. The adsorbent was washed in turn in 0.5 N hydrochloric acid, water, 0.5 N sodium hydroxide, water and was then resuspended in 7 M urea and HCl added until the pH was 3.0.

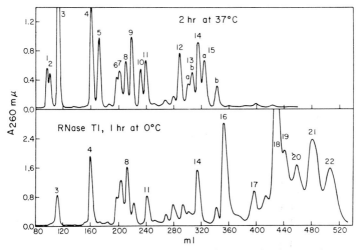

Fig. 2.1. Chromatographic patterns of a complete (upper half) and partial (lower half) T_1-RNase digest of yeast alanine tRNA on a 240×0.35 cm neutral DEAE-cellulose column in 7 M urea, 0.02 M Tris-chloride pH 8.0 using a linear NaCl gradient from 0-0.3 M. Notice the sharp symmetrical peaks without any noticeable trailing even for the longer oligonucleotides. (From Apgar et al. 1966.)

It was then thoroughly washed with 7 M urea adjusted to pH 3.0 with HCl. Although it is difficult to generalise about the conditions required, a column 30×1 cm may be tried at room temperature, eluting with 500 ml each of 0 and 0.25 M NaCl in 7 M urea (pH 3.0). DEAE-cellulose may be used instead of DEAE-Sephadex in this procedure (although the latter has a better flow rate) and some authors also use 7 M urea in 0.1 M formic acid instead of in HCl at pH 3.0. Fractionation at different pH's (e.g. from 2.5 to 4.0) may be used in an attempt to separate mixtures which do not easily resolve at pH 3.0. Longer columns will undoubtedly give better resolution than can be obtained on a 30-cm column.

2.3.2. Complete pancreatic ribonuclease digests

Chromatography on DEAE-cellulose (or DEAE-Sephadex) in the absence of urea, using an ammonium carbonate gradient at pH 8.5 is commonly used, because of the excellent resolution of small oligonucleotides. However, both the resolution and the yield of the larger oligonucleotides is poor and for example a heptanucleotide may be eluted from the column only with difficulty or by using 1 M sodium chloride and 7 M urea. Moreover, ammonium carbonate is not very volatile so that if a volatile salt is preferred triethylamine carbonate should be used. The following alternative procedure which was originally suggested by Rushizky et al. (1964) and used in the sequence studies on tyrosine RNA of yeast by Madison et al. (1966) is therefore proposed. This procedure may also be used to analyse T_1-RNase digests as an alternative to the method described in § 2.3.1. It involves chromatography on a short neutral-type DEAE-cellulose (or DEAE-Sephadex) column in 7 M urea which separates oligonucleotide according to chain length followed by re-running the various peaks separately on short, acid DEAE-cellulose columns in 7 M urea. In both cases the column dimensions were 25×0.35 cm. A linear gradient using 100 ml of each of 0 and 0.3 M NaCl in 7 M urea, 0.02 M Tris-chloride, pH 7.0, was used for the neutral column; and for the acid columns, 100 ml of 0 and 0.1 M NaCl in 7 M urea, 0.1 M formic acid. 7 M urea adjusted to pH 3.0 with HCl could also have been used, in which case the salt gradient would be run from 0 to 0.2 M NaCl, 7 M urea.

2.3.3. Partial enzyme digests

2.3.3.1. 'Half' molecules produced by partial T_1-RNase. Neutral pH DEAE-cellulose columns in 7 M urea are used – run either at room temperature or at approximately 55 °C, in a water-jacketed column. Thus Penswick and Holley (1965) used a column 100×0.5 cm and eluted with a linear salt gradient formed from 200 ml of each of 0.0, 0.1, 0.2, 0.3, 0.4 and 0.5 M sodium chloride in 7 M urea, containing 0.02 M Tris-chloride, pH 8.0 in six chambers of a Varigrad (see Peterson, this series, vol. 2, part II). The column was run at room temperature and 3.2-ml fractions were collected. Partial separation of the two 'halves' of alanine tRNA was obtained as shown in fig. 2.2. Peaks I and II were purified further by re-chromatography as before after diluting with an

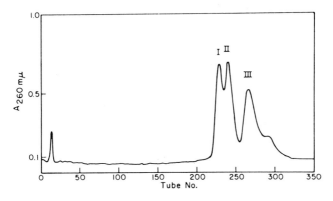

Fig. 2.2. Separation of two halves (peaks I and II) from intact (peak III) yeast alanine tRNA after very limited T_1-RNase digestion (§ 2.2.4.1) by chromatography on a neutral DEAE-cellulose column (100×0.5 cm) in 7 M urea at room temperature using a gradient of 0-0.5 M sodium chloride. (From Penswick and Holley 1965.)

equal volume of water. Each peak was then over 90% pure. Madison and Kung (1967) used, instead, a hot DEAE-cellulose column, 120×0.35 cm and eluted with a linear gradient of sodium-chloride in 7 M urea, 0.02 M Tris-chloride, pH 7.0 (volume not stated) at 55 °C. Fig. 2.3 shows the result which gave the 'halves' (the first 2 peaks) in sufficient purity for an unambiguous analysis. Note that when running hot columns it is necessary to remove air from all solutions used, otherwise

Subject index p. 261

air bubbles form as the temperature rises and are trapped in the column. The solution to be degassed is stirred by means of a magnetic stirrer in a Buchner flask, the inlet being blocked with a tight fitting rubber bung, and the outlet being connected to a vacuum (water) pump.

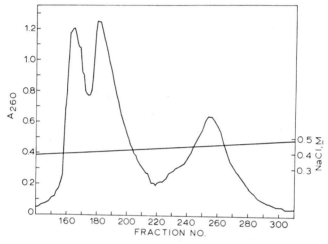

Fig. 2.3. Limited T_1-RNase digestion of yeast tyrosine tRNA fractionated by chromatography on a DEAE-cellulose column (120×0.35 cm) in 7 M urea (*p*H 7.0) at 55 °C and using a linear sodium chloride gradient. (From Madison and Kung 1967.)

2.3.3.2. Extensive partial T_1-RNase digests. This type of degradation produces the most complicated mixture of fragments likely to be encountered in sequencing low molecular weight RNA. Both partial products, which contain more than one G residue, and end-products of digestion are present. This means that fractionation on several columns may be required to purify a given product. However, under most conditions of digestion relatively few partial products are produced and are present in good yield so that this improves the chances of any one product being isolated in a pure form. The reason for this is the remarkable specificity of T_1-RNase for non-hydrogen-bonded guanine regions of the molecule; this point is discussed further in § 5.1. The following are the conditions used by Apgar et al. (1966). A partial digest of alanine tRNA was fractionated on a neutral DEAE-cellulose column (200×0.35 cm) exactly as described for a complete T_1-RNase digest (§ 2.3.1); a typical profile is shown in fig. 2.1 (lower section). A comparison with the upper part of fig. 2.1 shows that peaks 16-22 are not present in the complete T_1-RNase digest and are therefore partial digestion products. These were separately purified by rechromatography at 55 °C on 200×0.35 cm columns of DEAE-cellulose packed in 0.2 M NaCl, 7 M urea, 0.02 M Tris-chloride, pH 8.0 and eluted with a linear gradient formed from 150 ml each of 0.2 M and 0.4 M sodium chloride in 7 M urea-0.02 M Tris-chloride, pH 8.0. The flow rate was approximately 20 ml/hr. The usual analytical procedure, after desalting (see § 2.4 below), is to degrade material in each peak to completion with T_1-RNase (see § 2.2.1) and fractionate the products on the same neutral column. Peaks produced may be identified by position and ultraviolet spectrum and identification confirmed by further analysis (*e.g.* alkaline hydrolysis, or P-RNase digestion). If there is sufficient material, the partial fragment should also be degraded to completion with P-RNase and the products separated as under § 2.3.2 and identified either by chromatographic position or subsequent analysis. For example, peak 18 of fig. 2.1 was heterogeneous and was separated into three by re-chromatography at 55 °C (fig. 2.4). Subsequent digestion of each

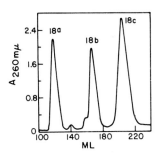

Fig. 2.4. Re-chromatography of peak 18 (fig. 2.1, lower half) on a hot (55 °C) DEAE-cellulose column (210×0.35 cm) using a linear gradient of 150 ml each of 0.2 and 0.4 M NaCl in 7 M urea–0.02 M Tris-chloride, pH 8.0. (From Apgar et al. 1965.)

of these three large fragments 18a, 18b, 18c with T_1-RNase followed by chromatography of the digests gave the results for 18a and 18b shown in fig. 2.5. The numbers 1, 3, 5, 10, 15 refer to end-products of T_1-

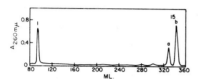
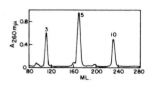

Fig. 2.5. Complete T_1-RNase digestion of fragments 18a and 18b (of fig. 2.4) fractionated on a neutral DEAE-cellulose 7 M urea column as in fig. 2.1. (From Apgar et al. 1965.)

RNase digestion, already characterised (fig. 2.1). 15a has a terminal 2′,3′ cyclic phosphate whilst 15b has a 3′-phosphate. The yield (§ 2.5) of the fragments 1 (a dinucleotide) and 15a plus b (an octanucleotide) was identical. Other fragments were treated in the same way, although not in all cases was purity achieved. Madison and Kung (1967) used exactly the same method to purify fragments – neutral followed by hot DEAE-cellulose columns. However, Dutting et al. (1966) preferred chromatography on acid *DEAE-Sephadex* columns at pH 3.0 in 7 M urea after the initial fractionation on neutral *DEAE-cellulose*. Both groups used long narrow columns, approximately 200×0.5 cm, and

---

Fig. 2.6. Isolation and sequence of some partial T_1-RNase products illustrating the type of column used in analysing yeast serine tRNA (from Dutting et al. 1966). (a) is a neutral (in 0.02 M Tris-chloride pH 7.5) DEAE-cellulose column in 7 M urea (approx. 0.8×210 cm). (b) and (c) show the further purification of regions from the first neutral column on an acid DEAE-Sephadex A25 column (approx. 0.7×210 cm) in 7 M urea adjusted to pH 3.0 with HCl. (d) is a further purification of the major peak in (c) using a hot (55 °C) DEAE-cellulose chromatography in 7 M urea and at neutral pH (0.02 M Tris-chloride pH 7.4 on a 0.7×210 cm column). (e) shows the analysis of a complete pancreatic RNase digest of fragment T-r in (d) on a DEAE-cellulose column using an ammonium bicarbonate gradient (see text). The products were characterised by a variety of techniques including their UV spectra, further degradation with enzymes, or by treatment with alkaline phosphatase follwed by paper electrophoresis at pH 2.7.

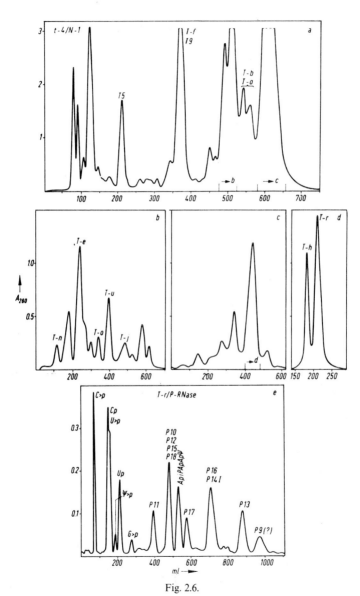

Fig. 2.6.

although the conditions varied slightly between experiments, a sodium chloride gradient from 0.05 M to 0.45 M NaCl with a total eluting volume of 1200 ml was commonly used. Most of the peaks eluting from the acid DEAE-cellulose (or DEAE-Sephadex) columns were homogeneous. Dutting et al. (1966) used a third column fractionation, the hot DEAE-cellulose column described above and eluted with 400 ml each of 0.2 to 0.5 M NaCl, to resolve peaks which were still impure. Such a procedure, shown in fig. 2.6, is the preferred method for studying partial digests. If there is a limited amount of material the most suitable procedure is a hot DEAE-cellulose column, followed by an acid DEAE-cellulose column. Recoveries of nucleotide from the acid columns are usually in the range of 80-90%.

2.3.3.3. Partial P-RNase digests. The separation methods are the same as those used for partial T_1-RNase digests described above. In summary this involves chromatography on neutral DEAE-cellulose, followed by acid DEAE-cellulose (or DEAE-Sephadex) chromatography in 7 M urea. Chang and RajBhandary (1968) used descending paper chromatography on Whatman No. 3 MM paper as an alternative second step in the purification of the *smaller* (up to decanucleotides) partial P-RNase digestion products from phenylalanine tRNA of yeast. In general, however, oligonucleotides have very low mobilities and chromatography for 72 hr may be necessary. The following solvents have been used: isobutyric acid-concentrated ammonia-water (66:1:33, v/v) at *p*H 3.7; *n*-propanol-concentrated ammonia-water (55:1:35, v/v).

The analysis of the products of a partial P-RNase digestion is identical to the analysis of the products of a partial T_1- RNase digestion (§ 2.3.3.2) and involves their further degradation with either T_1 or P-RNase to completion. Instead, and if yields permit, a partial product may be retreated with T_1- or P-RNase again using *partial* conditions. In this way Dutting et al. (1966) were able to obtain a more random degradation of serine tRNA than was possible in a *single* partial digestion. Thus partial products were found which were not present in the first partial digest. This second set of products was analysed by complete T_1- and P-RNase digestion and provided information necessary for the logical deduction of the sequence of serine tRNA. The conditions for the second partial T_1-RNase digestion were very similar to those of Apgar et al. (1966) described in § 2.2.4.1. About 20 mg of the partial product was used and the digests were fractionated on a hot DEAE-cellulose column followed by an acid DEAE-Sephadex column. For the second partial P-RNase digestion, approximately 1 part of enzyme for 100 parts of partial product was used. Digestion was for 20 min at 0 °C in 0.1 M Tris-acetate, pH 7.5, 0.01 M magnesium acetate; finally P-RNase was removed as described in § 2.2.4.2.

2.4. Desalting of nucleotide peaks from columns

This is necessary for subsequent analysis of oligonucleotides and several alternative methods include gel filtration, adsorption onto DEAE-cellulose, and dialysis. Dialysis is restricted to oligonucleotides of chain length about 8 or more, whilst the other two methods are generally applicable. Gel filtration is probably the method of choice because recoveries can be quantitative. Most authors, however, have used the second method. The technique is described by Fischer (1969) in vol. 1 of this series.

2.4.1. Gel filtration

A highly cross-linked dextran (e.g. G-10 Sephadex – see appendix 2) or polyacrylamide gel (e.g. Biogel P-2 – see appendix 2) must be used so that the small oligonucleotides are excluded from the gel (Uziel 1967). Sephadex is probably slightly easier to handle than the Biogel. G-10 Sephadex may be used for all except mononucleotides which should be put through the procedure described in § 2.4.2 below. G-25 Sephadex may be used if the oligonucleotide has a chain length of 5 or more. The procedure for preparing columns of Sephadex and polyacrylamide gels suitable for the purpose is described in a separate monograph of this series by Matthews and Gould (in prep.). Only a brief description is given here. The gel is swollen in water overnight and fines are removed by settling in a large volume of water and decanting the more slowly sedimenting particles. Columns with a diameter of not less than 2 cm and a volume of at least 10 times the volume of the sample to be desalted, are packed in sections under gravity. A dilute slurry of gel is poured into a column of water so that any air bubbles trapped in the slurry can escape. When packed, the column is washed with two volumes of distilled water before using. The sample is applied, washed in, and the effluent monitored at 260 mμ by an ultraviolet recorder (e.g. Uvicord) and fractions collected. The absorbancy peak is then pooled and concentrated to dryness by rotary evaporation in vacuo at 30-37 °C. At least 2 bed volumes of water are passed through the column to regenerate the gel.

2.4.2. Adsorption onto DEAE-cellulose

The sample is adsorbed at low ionic strength onto a small column of DEAE-cellulose which is then washed thoroughly with a very dilute volatile buffer to remove salts and urea. The oligonucleotides are eluted with a high concentration of the volatile salt, which is then removed under vacuum (Rushizky and Sober 1962).

DEAE-cellulose is washed as described (§ 2.3.1) and is equilibrated with 0.01 M triethylamine carbonate, pH 8.5 (see § 4.4 for preparation of triethylamine carbonate) or with 0.01 M ammonium bicarbonate, pH 8.5. A column is packed with a volume of absorbent – at least 2 ml/mg of oligonucleotide to be applied. The sample is diluted five times with the 0.01 M triethylamine carbonate (or ammonium bicarbonate) and applied to the column, which is then washed with 10 volumes of buffer or until the conductivity of the effluent is the same as the inflowing buffer. Oligonucleotides are then eluted with 2 M triethylamine carbonate or ammonium carbonate (a fairly large volume may be required) and *all* fractions with absorbancy at 260 mμ are pooled and concentrated to dryness by rotary evaporation at 30 °C in vacuo. The sample is re-dissolved in water and again concentrated to dryness. This procedure is repeated until no salt (whitish, crystalline appearance) remains. Triethylamine carbonate is more volatile than ammonium bicarbonate and is therefore preferable. If the latter is used, removal of the last traces of ammonium bicarbonate can be effected by mixing an aqueous solution of the oligonucleotide with Dowex-50 (pyridinium form) resin. The resin is filtered off and washed with dilute ammonium hydroxide. The filtrate and washings are evaporated to dryness twice from dilute ammonium hydroxide (RajBhandary et al. 1968).

2.4.3. Dialysis

This is only suitable if the oligonucleotide has a chain length of 8 residues or more. It is therefore generally only used for the desalting of partial digestion products. Dialysis tubing must be treated with ethylenediamine tetraacetic acid (EDTA) to remove metal ions, and boiled to destroy or leach out ribonucleases. Dialysis tubing (Visking size 20) is immersed for 24 hr in 0.01 M Na_2-EDTA $pH \sim 7.0$ and then transferred to distilled water and boiled for 15 min in three changes of distilled water before use. Tubing is tested for holes before use and is always handled with polythene gloves because fingers contain ribonucleases. The sample to be dialysed should half fill the tubing to allow for expansion as water enters under osmosis. Samples may be concentrated by freeze-drying or rotary evaporation after dialysis against three changes of water.

2.5. Yields of isolated oligonucleotides

The original absorbancy profile, for example that shown in fig. 2.1, is used to calculate the yields of the fragments. Where a single peak is known to be heterogeneous, the ratio of the several components must be estimated after further separation. The percentage of the total absorbancy at 260 mμ in each peak, or shoulder of a peak, may then be calculated relative to the total absorbing material eluted. The *relative molar yields* of each fragment may then be obtained calculated by dividing the observed percentage yields by ε (mM extinction coefficient) for the oligonucleotide. Values of ε for oligonucleotides may be found in the literature (Staehelin 1961; Zachau et al. 1966), or alternatively, the extinction coefficients of the constituent mononucleotides (appendix 4) may be summated to give a value for the oligonucleotide. (It will be noted that the composition of an oligonucleotide must be known before a yield can be calculated.) Figures may be corrected for hypochromicity (the fact that oligonucleotides have a lower absorbancy than the sum of the constituent mononucleotides). Hypochromicity is about 5% for dinucleotides and up to 30% for RNA and may be measured experimentally in a spectrophotometer after alkaline hydrolysis (and neutralisation) or after total enzymatic hydrolysis, e.g. by T_2-RNase (see below). The *relative molar yields* may be converted to actual *molar yields* if in addition to knowing the ε-values for the oligonucleotides the exact $\varepsilon_{260\ m\mu}$-value of the RNA is known. This latter value may also be calculated if the number of residues and hypochromicity is known. However, the yield is often best left as, and quoted in, moles relative to 1 mole of, say, a known dinucleotide.

2.6. Sequence in oligonucleotides

An outline of sequence analysis of the purified products of complete T_1- and P-RNase digestion can be briefly considered before the detailed practical procedures are given. There are many methods, both chemical and enzymatic, which are suitable for the degradation of oligonucleotides; these are summarised in table 2.1 which includes procedures developed in the radioactive studies described in chs.4-6. It will be noted that enzymatic digestion procedures are used extensively. Mononucleotides and mononucleosides are the smallest products and are identified by their electrophoretic and chromatographic properties and also by their ultraviolet absorption spectra. If previously unknown nucleotides are encountered they are identified and characterised as mononucleotides. Once a new nucleotide is recognised, even if its structure is unknown, it should no longer interfere with the sequence analysis and may in fact be useful in establishing overlaps between certain sequences. *Dinucleotides* are sequenced by hydrolysis with alkali or with enzymes such as T_2-RNase. Because of the specificity of the RNase used to isolate the dinucleotide its 3'-terminus is known, so that, for example, a dinucleotide from a T_1-RNase digest giving Cp and Gp must be CpGp (or CG). If it came from a P-RNase digest, it would be GpCp (or GC). *Trinucleotides* may in some cases be sequenced by composition alone, e.g. CCG, but in general a 5'-

TABLE 2.1

Procedures for sequencing end-products of T_1- and P-RNase digestion. (For rules regarding nomenclature, see appendix 1.)

Procedures	Specificity (arrow shows cleavage)	Products
Alkaline hydrolysis	↓ N-p-N	Mononucleoside 2'- and 3'-phosphates (via 2', 3'-cyclic phosphates)
T_2-RNase	↓ N-p-N	3'-mononucleotides (via 2', 3'-cyclic phosphates)
P-RNase	↓ Py-p-N	
P-RNase (after CMCT blocking, see § 6.5)	↓ C-p-N	Nucleoside 3'-phosphates (via 2', 3'-cyclic phosphates)
T_1-RNase	↓ G-p-N	
U_2-RNase	↓ Pu-p-N	
Spleen acid RNase	↓ N*-p-N	
Spleen exonuclease (phosphodiesterase)	Exonuclease from 5'-end	Mononucleoside 3'-phosphates
Snake venom phosphodiesterase	Exonuclease from 3'-end**	1. Mononucleoside (derived from 5'-end) 2. Mononucleoside-5'-phosphates 3. pNp (derived from 3'-phosphate end)
Polynucleotide phosphorylase	Exonuclease from 3'-end	1. Mononucleoside 5'-diphosphate 2. NpN and some NpNpN (from 5'-end)

(*continued*)

TABLE 2.1 (*continued*)

Procedures	Specificity (arrow shows cleavage)	Products
Micrococcal nuclease	↓ N-p-N	Mono-, and dinucleoside 3'-phosphates
Secondary splitting (see § 5.2.1)	↓ Py-p-A	Oligonucleoside 3'-phosphates

* Some preference for A residues.
** If 3'-phosphate present, enzyme acts very slowly.

terminal end-group procedure is required in addition to alkaline hydrolysis. Degradation with snake venom phosphodiesterase gives a nucleoside from the 5'-terminus, a 5'-mononucleotide from the middle nucleotide, and a nucleoside 3', 5'-diphosphate from the 3'-end. Thus CUG will give C, pU and pGp. For the sequence of *tetranucleotides*, further digestion with P-RNase (specific for pyrimidines), or U_2-RNase (specific for purines) will allow identification of all except CUCG from CCUG and UCUG from UUCG. This point is illustrated in table 2.2, taken from Holley (1968). Thus for these four tetranucleotides and larger oligonucleotides, other approaches are needed.

The problem was solved by digestion with snake venom phosphodiesterase – an exonuclease attacking from the 3'-end – under conditions in which the degradation to 5'-mononucleotides was incomplete (Holley et al. 1964). Thus a series of intermediate products was isolated and, by analysing these, the authors were able to deduce the sequence of an octanucleotide. To illustrate this method, consider the pentanucleotide U-C-U-C-G isolated from a combined T_1-RNase and alkaline phosphatase digest. On partial digestion the following products should be obtained: U-C-U-C, U-C-U and U-C, in addition to the 5'-monucleotides released. The analysis of this series of products by alkaline hydrolysis allows one to reconstruct the pentanucleotide

TABLE 2.2
Information required for the identification of tetranucleotides.

Information required	Tetranucleotides found in a pancreatic RNase digest		Tetranucleotides found in a RNase T_1-digest		
Nucleotide composition	AAAC GGGC	AAAU GGGU	AAAG	CCCG	UUUG
Plus identity of 5′-terminal	AGGC GAAC	AGGU GAAU	ACCG AUUG	CAAG CUUG	UAAG UCCG
Composition, 5′-terminal, plus analysis of digest with other RNase*	AGAC ⎫ AAGC ⎭ GAGC ⎫ GGAC ⎭	AGAU ⎫ AAGU ⎭ GAGU ⎫ GGAU ⎭	ACAG ⎫ AACG ⎭ AUAG ⎫ AAUG ⎭	CACG ⎫ CCAG ⎭ CAUG ⎫ CUAG ⎭ UACG ⎫ UCAG ⎭	ACUG ⎫ AUCG ⎭ UAUG ⎫ UUAG ⎭
All of above insufficient			CUCG ⎫ CCUG ⎭		UCUG ⎫ UUCG ⎭

* The pairs of tetranucleotides in brackets have identical nucleotide compositions and identical 5′-terminal residues.

sequence as each product will differ from the next by a single mononucleotide. Partial digestion with spleen phosphodiesterase may be used in a similar way but has only been used extensively in studying ^{32}P-labelled oligonucleotides (Sanger et al. 1965). This method is, however, very useful for sequencing the longer P-RNase products (Min-Jou and Fiers 1969). Micrococcal nuclease is another enzyme which has been used in specific cases for sequencing *tetranucleotides*. Thus CCUG will give CC and UG in good yields. Other procedures, e.g. partial digestion with pancreatic ribonuclease and partial digestion with polynucleotide phosphorylase, have been used to sequence oligonucleotides in specific instances, but have not been generally used.

The usual practical procedure in sequencing oligonucleotides may thus be summarised. A T_1-RNase digest is prepared, the products are

Subject index p. 261

fractionated and all the di-, tri- and tetranucleotides are sequenced. Larger nucleotides or any difficult tetranucleotides are sequenced by partial digestion with venom phosphodiesterase after the terminal phosphate groups have been removed by treatment with alkaline phosphatase. Alternatively, larger oligonucleotides may be isolated as dephosphorylated products after combined digestion with T_1-RNase and alkaline phosphatase (§ 2.2.2). A P-RNase digest is then prepared and the products sequenced. As the products are on average smaller than in the T_1-RNase digest, their sequence determination is usually fairly easy and is achieved by the same methods as for the T_1-products. Occasionally, extremely large nucleotides may be isolated in the T_1-digest for which partial digestion does not work well, or for which confirmatory evidence is required. Other procedures suitable for such difficult cases will be discussed in the chapter on the sequence of radioactive oligonucleotides. The yields of oligonucleotides must also be measured (§ 2.5).

2.6.1. Sequence methods

In the following list of methods used in sequence analysis, cross reference should be made to ch. 3 for any details of high-voltage electrophoresis on paper and to chs. 4 and 5 for further details of procedure and especially for the interpretation of typical results and any difficulties that may be encountered. Although ch. 4 describes results for radioactive RNA, the methods and the interpretation are more or less the same as for unlabelled RNA. Even with unlabelled oligonucleotides, paper techniques can be extensively used in sequence analysis of oligonucleotides because of their simplicity and ease of operation compared to column chromatography. Therefore digestions are often carried out in small volumes and in low salt concentrations, and the whole mixture is applied to the paper for fractionation at the end of the incubation. The amount of oligonucleotide degraded in any one procedure should be calculated as that amount which will give 0.5-1.0 A_{260}-units (20-40 μg) of nucleoside or mononucleotide at the final analytical step, assuming that this is to be paper chromatography. This is higher than the smallest amount of mononucleotide that can be detected by ultraviolet absorption on paper (about 0.1 A_{260}-units or approx. 10 nmoles) so as to allow for elution of the material and accurate quantitation of the results. By using thin-layer chromatographic systems, the sensitivity may be further increased (Randerath 1967; Rogg and Staehelin 1969). Thus, Randerath and Randerath (1967) made quantitative measurements on amounts of nucleoside as low as 1 nmole and even lower amounts than this could be detected. Uziel et al. (1968) have also developed a very sensitive small-scale column procedure for analysing nucleosides which they, also, claim to be sensitive to 1 nmole of nucleoside. As will be seen later, sensitivity can be increased nearly 1000-fold by using ^{32}P and where this is possible paper and thin-layer fractionation methods offer great economy in time and material. In all enzymic digestions the optimum conditions may vary slightly because of variations in activity of enzymes and possibly for other reasons. Therefore, trial experiments should be carried out with either variations in the time of digestion, or in the amount of enzyme, using the conditions suggested below only as a guide in order to establish the optimum conditions. With partial digestions, as in §§ 2.6.1.6-2.6.1.8 below, different oligonucleotides will undoubtedly degrade at different rates, so that trial small-scale experiments should be performed on the actual oligonucleotide in question. Oligonucleotides must be dry and free from salt before analysis.

2.6.1.1. Alkaline hydrolysis. 50 µl of 0.3 N KOH is added to the dried oligonucleotide which is then incubated in a stoppered tube or a capillary tube sealed at each end for 18 hr at 37 °C. Often authors use 0.5-1 M base for 18 hr at 37 °C, but this unnecessarily raises the salt concentration and is needed only if the RNA concentration is over 20 mg/ml (e.g. over 1 mg of nucleotide in the above procedure) or if appreciable buffer is present in the sample (Bock 1967). Zamir et al. (1965), and others, then apply the hydrolysate as a short band, or spot, to Schleicher and Schull No. 589 paper (Whatman No. 1 or No. 3 MM may also be used) and separate the mixture by two-dimensional ascending chromatography. The solvent system for the first dimension is propan-2-ol : water (60 : 40) with ammonia in the vapour phase accord-

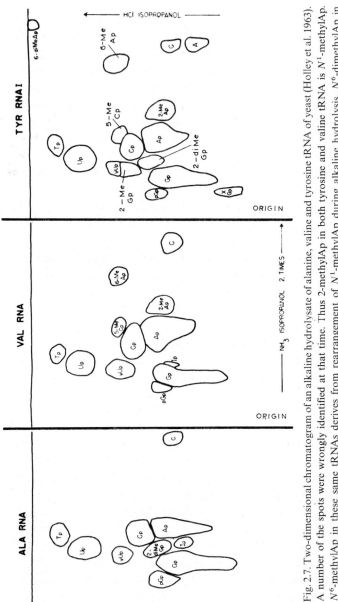

Fig. 2.7. Two-dimensional chromatogram of an alkaline hydrolysate of alanine, valine and tyrosine tRNA of yeast (Holley et al. 1963). A number of the spots were wrongly identified at that time. Thus 2-methylAp in both tyrosine and valine tRNA is N^1-methylAp. N^6-methylAp in these same tRNAs derives from rearrangement of N^1-methylAp during alkaline hydrolysis. N^6-dimethylAp in tyrosine tRNA has subsequently been shown to be N^6-isopentenylAp. XGp is an alkaline resistant dinucleotide, 2′-O-methylguanylylGp. N^1-methylGp (running in a similar position to N^2-methylGp) is present in alanine and valine tRNA but is not shown. 5,6-dihydroUp (in all three tRNA's) and N^1-methylIp (in alanine tRNA) were not detected as they degraded to non-ultraviolet-absorbing material during hydrolysis.

ing to Markham and Smith (1952). The chromatogram may be re-run after drying in order to improve the resolution of the nucleotides, some of which, but not all, have very low mobilities in this solvent. The chromatogram is then developed at right angles to the first direction using the propan-2-ol–HCl system of Wyatt (1951) (680 ml propan-2-ol and 176 ml concentrated HCl diluted to 1 litre with water). This system gives a good general separation of the four common mononucleotides and some of the minor mononucleotides, as well as of the nucleosides and nucleoside 3′, 5′-diphosphates (see below). Fig. 2.7 is an example of this fractionation applied to alkaline hydrolysates of three purified tRNA molecules from yeast. R_f's cannot be given with any accuracy as the first dimension was run twice (however, see table 7.2) Notice the 'tailing' of the mononucleotides, e.g. Gp and Ap, which is probably due to over-loading.

An alternative fractionation system for mononucleotides is high-voltage paper electrophoresis at pH 3.5, described first by Markham and Smith (1952). This gives better separation of the four major mononucleotides and nucleosides from each other (see table 7.2) but does not give such good resolution of many of the minor nucleotides which often have an aliphatic substituent (e.g. a methyl group) on the ring which has little or no effect on the pK_a of the base. Therefore, this electrophoretic separation is best combined with one or other of the paper-chromatographic methods as a second dimension. This was the method of choice in the radioactive work on tRNA and further details of results using this method are given in ch. 7.

Besides the nucleoside 3′, 5′-diphosphates and nucleosides, derived from the 5′- and 3′-end, respectively, of for example tRNA molecules which have in general the form of $pG\text{-}(N)_n\text{-}C\text{-}C\text{-}A_{OH}$, the following other products may be expected: (i) breakdown products of nucleotides unstable to alkali (see table 2.3); (ii) dinucleotides of the form 2′-O-methyl NpNp where a methyl group on C-2 of the ribose ring prevents cleavage of the phosphodiester bond attached to the neighbouring C-3 position; (iii) dinucleotides such as ApAp which are rather slowly hydrolysed by alkali. The methylated derivative of this dinucleotide is even more slowly hydrolysed, thus $m_2^6A\text{-}m_2^6A\text{-}$ was still present after

TABLE 2.3
Mononucleotides unstable to alkaline hydrolysis in tRNA.

Nucleotide	Product of hydrolysis	Reference
1. 5,6-dihydro U	β-ureidopropionic acid N-ribotide	Madison and Holley (1965)
2. 2-thio-5-uridine acetic acid methyl ester	Corresponding acid	Baczynskyj et al. (1968)
3. Uridine-5-acetic acid ester	Corresponding acid	Gray and Lane (1968)
4. 3-methyl C*	3-methyl U	Hall (1963b)
5. N^4-acetyl C	C	Feldmann et al. (1966)
6. 1-methyl I*	5-amino-imidazole-4 (N-methyl)-carboxamide-1-ribotide	Hall (1963a)
7. 7-methyl G	4-amino-5-(N-methyl)-formamido isocytosine ribotide	Dunn (1963)
8. 1-methyl A*	N^6-methyl A	Dunn (1963)
9. N-(purin-6-ylcarbamoyl)-threonine ribotide		Schweizer et al. (1968)

* Conversion to product shown may not be complete if mild alkaline hydrolysis is used.

90 hr alkaline hydrolysis with 1 M NaOH at room temperature (Nichols and Lane 1966).

2.6.1.2. *T_2-RNase* (Rushizky and Sober 1963). This enzyme, from the fungus *Aspergillus*, cleaves RNA giving mononucleoside 3'-phosphates. It is mainly used as an alternative to alkaline hydrolysis, when it is suspected that degradation of a nucleotide is occurring in alkali (see table 2.3). The dried oligonucleotide is digested with 0.1 to 1.0 units of T_2-RNase (appendix 2) per mg of oligonucleotide for 6 hr at 37 °C in 0.1 M ammonium acetate, pH 4.5 in a volume of 0.1 ml. The enzyme is less stable than T_1-RNase but is still very stable compared to most enzymes and an aqueous solution is completely stable at 2 °C. The products of digestion are fractionated as described for the products of alkaline hydrolysis (§ 2.6.1.1). Crude preparations of T_2-RNase (containing T_1-RNase, but only very low phosphatase activity) for example 'RNase CB' (appendix 2) may also be used. However, much higher amounts of enzyme, up to twice the weight of enzyme to RNA, may be required. A crude preparation of T_2-RNase may be simply prepared by heat treatment of 'Takadiastase powder A' purchased from Sankyo Co., Ltd. (appendix 2) as described by Hiramaru et al. (1966). T_2-RNase digestion, like alkaline hydrolysis, does not cause the

TABLE 2.4
Minor nucleotides in RNA identified since Hall (1965).

Nucleotide	Reference	Source (tRNA)
1. 4-thio U	Lipsett (1965)	*E. coli*
2. 5,6-dihydro U	Madison and Holley (1965)	All species
3. Uridine-5-acetic acid ester*	Gray and Lane (1968)	Yeast, wheat germ
4. 2-thio-5-uridine acetic acid methyl ester	Baczynskyj et al. (1968)	Yeast
5. 5-methylaminomethyl-2-thio U	Carbon et al. (1968)	*E. coli*
6. 2-thio C	Carbon et al. (1968)	*E. coli*
7. N^4-methyl-2'-O-methyl C	Nichols and Lane (1966)	*E. coli***
8. N^4-acetyl C	Feldmann et al. (1966)	
9. N^6-(γ,γ-dimethylallyl) A purine 6-(3-methylbut-2-enylamino)†	Zachau et al. (1966)	Yeast mammals
10. 6-N-(cis-4-hydroxy-3-methyl but-2-enylamino) purine	Hall et al. (1967)	Plant
11. 2-methylthio-N^6-(γ,γ-dimethylallyl) A	Burrows et al. (1968)	*E. coli*
12. N-(purin-6-ylcarbamoyl)-threonine ribotide	Schweizer et al. (1968)	Yeast, *E. coli*, mammals
13. Uridine-5-oxyacetic acid	Murao et al. (1970)	*E. coli*
14. Compound Y	Nakanishi et al. (1970)	Yeast

* Unknown ester.

** Found in 16S ribosomal RNA.

† Trivial name, isopentenyl A.

cleavage of the phosphodiester bond on C-3 where C-2 of the ribose ring is methylated. A 2'-O-methylated nucleotide is, therefore, released as a completely stable dinucleotide. Dinucleotides such as A-A* are also rather slowly cleaved even with purified T_2-RNase when the substituent denoted by an asterisk on A is an isopentenyl group (see nucleotides 9, 10 and 11 of table 2.4) so these may be expected unless digestion conditions are particularly vigorous (Dr. M. Gefter, personal communication).

2.6.1.3. Snake venom phosphodiesterase (Holley et al. 1965b). 10 µl of 1 M Tris-chloride, pH 7.5, 10 µl of 0.5 M magnesium chloride and 80 µl of snake venom phosphodiesterase (appendix 2) at 2 mg/ml and further purified according to Keller (1964) is added to the dried oligonucleotide and incubated for 4 hr at 37 °C in a stoppered tube or capillary tube. The digest is analysed by the two-dimensional paper chromatographic system as described in § 2.6.1.1, or by paper electrophoresis in 20% acetic acid adjusted to pH 2.7 with ammonia. The 5'-terminal residue is released as a nucleoside so that at pH's between 2.5 and 3.5 a separation is achieved between uridine (which is neutral) and guanosine, adenosine and cytidine (which have increasing mobilities towards the negative electrode). Nucleoside 3', 5'-diphosphates released from the 3'-end move faster than pC, pA and pG, although pU is rather similar to pCp and pAp. However, pU may be distinguished from pCp and pAp by paper chromatography on propan-2-ol–NH_3-water, where the diphosphates have much lower mobilities than the monophosphates. It should be noted that commercial venom phosphodiesterase has been used without further purification in radioactive studies (§ 4.4.2.4) although it is probably preferable to purify it.

2.6.1.4. Pancreatic RNase digestion of T_1-oligonucleotides and T_1-RNase digestion of P-RNase oligonucleotides. The expected products and method of fractionation are discussed under radioactive sequence methods (§§ 4.4.2.1, 4.4.2.2). For unlabelled material, ionophoresis using a non-ultraviolet-absorbing solvent is necessary (e.g. 5% or 10% acetic acid adjusted with ammonia to pH 3.5).

2.6.1.5. U_2-RNase digestion of T_1-end products. Under defined conditions this enzyme, extracted from the slime mould *Ustilago*, hence the terminology U (Arima et al. 1968), is specific for purine residues so that for T_1-oligonucleotides it will give products terminating in Ap and Gp. Thus any T_1-oligonucleotide containing a single A may be sequenced. For example, CACUG gives CA and CUG. The enzyme is also particularly useful for sequencing the 3'-terminal oligonucleotide of tRNA, e.g. $CACCA_{OH}$ of phenylalanine tRNA of *Torulopsis* yeast was split into CA and CCA_{OH}, thus defining its sequence. Suitable conditions are as follows: 0.1-1.0 units of U_2-RNase (appendix 2) is used per mg of oligonucleotide and incubation is for 22 hr at 37 °C in 0.05 M sodium acetate, 0.002 M EDTA containing 0.1 mg/ml bovine serum albumin at *p*H 4.5 in a volume of 0.1 ml. Suitable fractionation systems are described in the section under radioactive methods (§ 6.6).

2.6.1.6. Partial digestion with snake venom phosphodiesterase. This procedure was introduced by Holley et al. (1964) and is a general method for sequencing penta-, and larger, oligonucleotides. As a 3'-phosphate group inhibits the action of the enzyme, the oligonucleotide must first be treated with bacterial alkaline phosphatase (or it may be isolated by the combined treatment with T_1-RNase and alkaline phosphatase on the RNA). Holley et al. (1964) used the following method: 0.5 mg of oligonucleotide was incubated 1 hr at 37 °C in 0.1 ml of 0.1 M. Tris-chloride, *p*H 8.3 and 0.1 ml (0.4 mg/ml) bacterial alkaline phosphatase. The phosphatase was removed after diluting the solution with 0.6 ml water by three extractions with 0.8 ml of water-saturated phenol; phenol was removed from the aqueous layer by at least five extractions with ether which was then evaporated off. 0.1 M Tris-chloride, *p*H 7.5, 0.1 ml, 0.1 M magnesium chloride, 0.1 ml, and snake venom phosphodiesterase (about 10 μl of 2 mg/ml) were added and the mixture was incubated at room temperature for about 15 min. After the digestion 0.4 g of urea and 3 ml of 7 M urea was added and the solution loaded immediately onto a 0.35×30 cm column of DEAE-cellulose in 7 M urea which was eluted with a linear gradient using 60 ml of each of 0 and 0.6 M sodium acetate, *p*H 7.5 in 7 M urea.

Fig. 2.8 shows the degradation of A-U-U-C-C-G. The peaks were desalted, hydrolysed with 50 μl 0.5 KOH and applied directly onto paper and subjected to two-dimensional paper chromatography (§ 2.6.1.1). Mononucleotides and nucleosides were identified by position and ultraviolet spectrum.

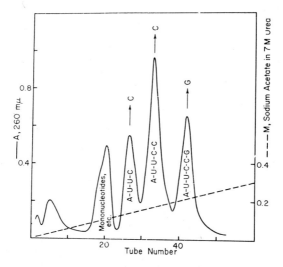

Fig. 2.8. Fractionation of a partial venom phosphodiesterase digest of the oligonucleotide A-U-U-C-C-G (from yeast alanine tRNA) on a DEAE-cellulose column (0.35 × 30 cm) in 7 M urea. The nucleoside released after alkaline hydrolysis is indicated above each peak. (From Holley 1968.)

This is probably the most valuable method available for the larger oligonucleotides and it is worth noting here a simple modification of this procedure which has been used in the radioactive studies. The phenol extraction step is omitted and no attempt is made to remove phosphatase before adding the venom phosphodiesterase. Mononucleotides are then converted to nucleosides whilst oligonucleotides lacking a terminal phosphate remain resistant to the phosphatase. This modification has the advantage that some of the smaller oligonucleotides are now no longer obscured by the mononucleotide peak (see fig. 2.8).

2.6.1.7. Micrococcal nuclease. Both mononucleoside 3'-phosphates and dinucleotides are produced as end-products of digestion with this nuclease and arise by a combination of endo, and exonuclease attack from the 3'-phosphate end. It is thus often possible to get a good yield of the dinucleotide from the 5'-end (Sulkowski and Laskowski 1962). The oligonucleotide may be digested as described in § 4.4.2.3 and the products fractionated either by paper electrophoresis, for example at pH 3.5 or by column chromatography on DEAE-cellulose in 7 M urea. Using this enzyme Zamir et al. (1965) isolated (T,ψ) from a digest of $(T,\psi,C)G$ thus allowing the sequence $(T,\psi)CG$ to be deduced. The conventions used will be noted by the reader. Thus an unknown sequence is enclosed within brackets with commas between symbols. Known sequences in contrast are not bracketed. Further examples appear in appendix 1.

2.6.1.8. Polynucleotide phosphorylase. This enzyme degrades oligonucleotides in the presence of inorganic phosphate in a stepwise fashion from the 3'-hydroxyl end and thus may be used as an alternative procedure to partial venom phosphodiesterase digestion (§ 2.6.1.6 above) Madison et al. (1967b) found that with polynucleotide phosphorylase (appendix 2) it is possible to get good yields of a trinucleoside diphosphate which, in contrast, is found in rather low yield in partial digests with venom phosphodiesterase. A suitable procedure is as follows: 0.2-0.5 mg of oligonucleotide is first dephosphorylated with 0.05 mg bacterial alkaline phosphatase in 0.15 ml 0.1 M Tris-chloride pH 8.0 plus a drop of chloroform for 1 hr at 37 °C. Then 0.08 mg of polynucleotide phosphorylase in 0.1 ml (available from Worthington Biochem. Corp., or further purified by the method of Singer and Guss 1962), 0.05 ml of 0.1 M potassium phosphate pH 7.0, and 0.1 ml 0.01 M magnesium chloride were added and incubation was continued for 2 hr at 37 °C. The sample was diluted with 5 ml of 7 M urea and fractionated and analysed, as described in § 2.6.1.6.

In summary, procedures 2.6.1.1-2.6.1.4 above are fairly standard methods and were described before any sequence analysis of pure RNA molecules was attempted. In these procedures, the interpretation of results is generally fairly straightforward, especially where the analysis of the ultraviolet spectrum of the mononucleotide is always the last step (§ 2.6.1.9). This leaves little room for error. The more sophisticated procedures (§§ 2.6.1.5, 2.6.1.6) are needed to establish the sequence of oligonucleotides (some tetranucleotides, pentanucleotides and larger). Here the interpretation of the results of an experiment has to be made with more caution and it is wise to have more than the minimum information necessary to deduce the sequence of an oligonucleotide, if at all possible. A few examples of the difficulties encountered and the methods used in establishing the sequence of two long T_1-oligonucleotides (17 and 21 residues in length) are given in ch. 6. The results of partial enzymatic digestion and how these are used to derive the total structure of an RNA will be discussed with reference to ^{32}P-5S RNA in ch. 5.

2.6.1.9. Spectral identification. Reference ultraviolet spectra for the four common and a few of the minor nucleosides and bases are quoted in Beaven et al. (1955) although a more recent compilation of spectra is by Venkstern and Baev (1965). Ultraviolet spectra and other data, such as chromatographic and electrophoretic mobility of more recently discovered minor nucleotides must be sought in the literature (see tables 2.3 and 2.4 for references). If a compound is found which has different mobilities and spectral properties from any of the known compounds its structure may be determined by high-resolution mass spectroscopy. For this purpose attempts should be made to isolate the free base if it is a purine derivative (by acid hydrolysis), or the nucleoside (by bacterial alkaline phosphatase treatment of the nucleotide) if a pyrimidine. Although the technique is very sensitive a reasonable quantity such as 50 μg of material should be prepared. The ultraviolet spectrum of a compound should be measured at pH 2.0 (in 0.01 N HCl) and at pH 12 (0.01 N KOH) against a paper 'blank' also containing the 0.01 N HCl or KOH. Bits of paper containing the nucleotide and 'blanks' (of equal area, or weight as the sample) are covered with water in small tubes and left overnight to allow the material to elute. Spectra of the dinucleotides are also often sufficiently characteristic to allow identification (Venkstern and Baev 1965). If the various ratios $A_{250/260}$, $A_{280/260}$, $A_{290/260}$ are plotted then the spectra of quite large oligonucleotides are distinctive enough to be able to use this feature as a means of recognising the presence of a particular oligonucleotide. This is important for example in the analysis of the end-products of a partial T_1- or P-RNase digestion product (see, e.g., Zachau et al. 1966).

CHAPTER 3

High-voltage paper electrophoresis

3.1. Introduction

Many of the techniques used in the sequence determination of nucleic acids are extremely simple and require no elaborate equipment. An adequate description of techniques such as paper chromatography (e.g. Markham 1957) and elution and ultraviolet detection of nucleotides (e.g. Heppel 1967) is available in the literature. Electrophoretic methods provide the basis of most of the fractionation methods developed by Sanger and his collaborators for working with radioactively-labelled nucleic acids, and a full description of the equipment and techniques is given here. There are also descriptions of the technique in recent articles by Smith (1967), Sanger and Brownlee (1967) and Ruschizky (1967). Applications of the methods appear in the succeeding chapters. For studying unlabelled RNA, electrophoresis is not essential but it is certainly a useful technique, for example in the purification of some of the smaller oligonucleotides that may elute as mixtures in column chromatography. It may also be used as an adjunct to paper chromatography in studying the sequence of oligonucleotides.

3.2. High-voltage electrophoresis

Paper electrophoresis of nucleotides was first described by Markham and Smith (1952). They immersed a piece of paper in chloroform in a museum jar, and applied a voltage gradient across the ends of the paper which hung outside over the edge of the jar. A more sophisticated apparatus, in which all of the paper and the electrode compartments were immersed in a coolant, was described by Michl (1951) and is similar to the 'hanging tank' described below. A slightly different arrangement was used by Katz et al. (1959) and modified by Naughton and Hagopian (1962); it is similar to the 'up-and-over' tank described below. An alternative arrangement for electrophoresis, in which the paper is cooled by contact with a cooled flat metal plate was originally proposed by Michl (1952). A modification of this (Gross 1961) is widely used, and is a useful and versatile piece of equipment as the pH may be easily changed. However, for many purposes, the organic solvent-cooled systems are more convenient to use and appear to give better separations than the flat-plate system. The latter is, therefore, not described here in detail.

3.3. The 'hanging' electrophoresis tank

In this system the sheet of paper, 47×57 cm, on which the separation occurs, is hung vertically, in a manner similar to descending chromatography, between a trough, which is one electrode compartment, and the bottom of the tank which is the other electrode compartment. The arrangement is illustrated in figs. 3.1 and 3.2, with the various components of the tank, the lid, the cooling-coil and the electrodes, drawn separately in the succeeding figures (figs. 3.2-3.5). The detailed design of this tank in the form described here was due to Dr. R. Ambler, University of Edinburgh, King's Building, Edinburgh, Scotland, to whom I am grateful for permission to reproduce this work. Heat formed during electrophoresis is dissipated into the organic solvent coolant, in which the paper is completely immersed. The coolant, in turn, is cooled by water circulating through the glass or stainless-steel cooling coil. The preferred organic solvent is white spirit (Varsol, a petroleum fraction containing dodecane and similar hydrocarbons). This is cheap, non-toxic, and has a fairly high flash point. It also does not affect PVC or Perspex, and does not leave any ultraviolet absorbing residue. This tank is convenient to use and, for example, is the tank which is used for the separation of the four mononucleotides in an alkaline hydrolysate of RNA at *p*H 3.5 on paper. Its disadvantages are that the length of paper is a constant (57 cm) and it also cannot be used with fragile papers, of low wet strength such as DEAE-cellulose. No such tanks are available commercially to the author's knowledge, although they may be assembled from parts which are available (see legends to figs. 3.2–3.5).

3.4. 'Up-and-over' tanks

In this system the paper is supported on a Perspex rack which can be inserted into the electrophoresis tank over a central partition which separates the electrode compartments. Several commercial forms of this apparatus are available and one made by Savant Instruments, U.S.A. (appendix 3) made of Lucite (polymethacrylate) may be recommended. A rather similar apparatus, but moulded from fibre glass, may be bought from Gilson Medical Electronics (appendix 3) but has the disadvantage that it is opaque. The design of tank used in the Laboratory of Molecular Biology, Cambridge, is shown in fig. 3.6. It is made from 1-inch clear Perspex which is glued at all joints and also screwed at 2-inch intervals. The Perspex rack is designed to take either the shorter (57 cm) or longer (85-90 cm) lengths of paper. As there must be no electrical leak across the central partition of the tank, the joining of the partition to the base of the tank is particularly critical. An improvement in design has been suggested by Dr. C. Milstein, which involves having a double partition separated by a gap into which glue or dissolved Perspex is poured. The dimensions of the tank shown in fig. 3.6 and the Perspex rack would have to be altered only slightly to accommodate this modification. A similar arrangement is suggested by Ruschizky (1967). A simpler type of 'up-and-over' tank may be assembled using a Shandon, Panglass Chromatank, Model 500 as was used for the 'hanging' tanks. In this system there is a large glass

---

Fig. 3.1. View of upper half of the 'hanging' tank. Note the attachment of the lid to the tank, and the position of the trough forming the upper electrode compartment. For other details refer to figs. 3.2-3.5 and the text.

Fig. 3.1.

trough supported on glass feet on the bottom of the tank holding at least 200 ml buffer, which is one electrode compartment. The buffer outside this trough forms the other electrode compartment. It requires a modification of the Perspex rack of fig. 3.6 to allow this to be used to support the paper with one end of the paper inside the trough and the other end outside it. This modification, however, makes the tank rather less convenient to use than if there is a partition; also the width of paper which can be used must be reduced from 45 cm to at least 44 cm as there must be room for the rack to drop down the side between the walls of the tank and the trough. Nevertheless, as all components (two long electrodes, cooling coil, lid, large glass trough) are available from T.W. Wingent, Ltd., this tank may be easily assembled (see appendix 3).

3.5. Power supply and safety precautions

For paper electrophoresis it is convenient to be able to apply up to 5 kV and although many of the papers used (e.g. Whatman No. 540, and cellulose acetate) absorb low volumes of buffer and take low currents, the DEAE-cellulose electrophoresis conducts a large current even at rather low voltages. For example, using the standard 7% formic acid system, a 45×85 cm paper takes approx. 140 mA at 1 kV. Therefore, the minimum requirements for both paper and DEAE-

---

Fig. 3.2. A side view of a 'hanging' tank electrophoresis system (see also fig. 3.1). The moulded glass tank (d) – a Panglass Chromatank (Model 500 from Shandon Scientific Co., 65 Pound Lane, Willesden, London) is attached to a piece of bakelite (b), hinged to the bakelite lid (fig. 3.3), over a protecting strip of rubber (c) by means of a metal strip (a). The cooling coil (in fig. 3.4) is tied with twine to the lid. The upper electrode compartment in a glass trough (e) is held at each end by two glass rods from a lip on the side of the tank. The electrophoresis paper (k) is held in the upper electrode compartment by means of a glass rod and hangs over another glass rod (i) (held in position by a piece of bakelite (h) at each end from the other lip on the tank) into the lower electrode compartment (g). The tank is completely filled (f) with 'white spirit 100', or Varsol (from Esso Petroleum Co.). The electrodes enter at the back of the tank, one dipping into the upper trough and the other (j) passing down the back of the tank to the lower compartment. All parts (figs. 3.3-3.5) other than the tank itself, are available from T. W. Wingent, Ltd. (appendix 3).

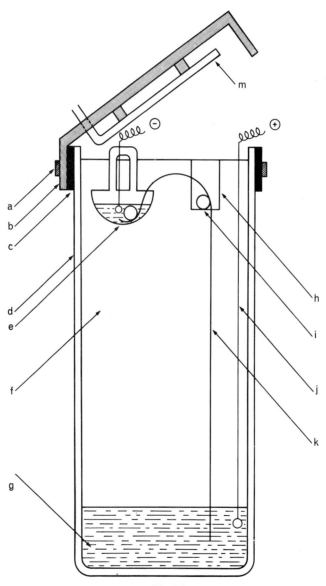

Fig. 3.2.

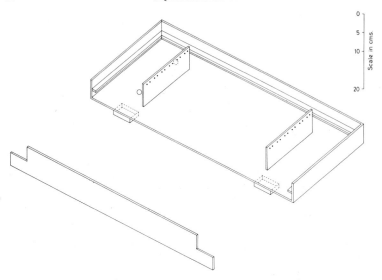

Fig. 3.3. Bakelite lid for the 'hanging' tank. All joints are screwed. The lid, itself, is attached by hinges to the strip of bakelite which is held on to the tank as shown in fig. 3.1.

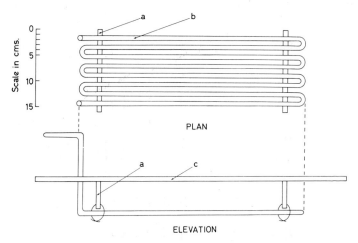

Fig. 3.4. Plan and elevation of glass cooling coil for 'hanging' tank. The glass coil (b) is held with twine onto the vertical member (a) of the bakelite lid (c).

cellulose electrophoresis are a power-pack delivering 5 kV and 250 mA. One providing 5 kV and 500 mA is preferable if two concurrent runs at the same voltage are needed. Brandenburg, Ltd. and Locarte, Ltd.

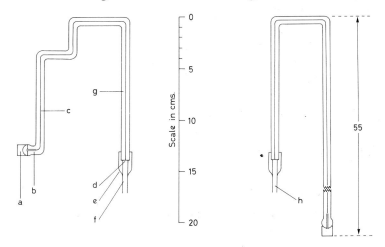

Fig. 3.5. Long and short electrodes for the 'hanging' tank. A platinum electrode (a) is connected to a nickel alloy wire (c) running through a glass tube (g). At (b) the alloy is fused to the glass. Outside the tank at (d) the wire is connected to a fully insulated wire (f) capable of taking the required load. This is connected to the neutral ($-$ve) output. The joint is protected by a glass sheath (e). The longer electrode (right of figure) is connected to the live ($+$ve) output (h).

(appendix 3) manufacture suitable equipment. In using such equipment, the operator must be protected from inadvertently opening the tank without isolating the power supply, because an electric shock at this voltage would probably be lethal. The tank, therefore, is isolated in a fume cupboard with a micro-switch located so that opening the hood of the fume cupboard automatically switches off the current. An operator should be aware that the current must be turned off manually before opening the fume cupboard, as the micro-switches are designed as a safety device should this precaution be omitted. Another hazard of this equipment is the fire risk should the 'white spirit' ignite. Detectors must, therefore, be installed in the fume cupboards which activate a CO_2-injection system should fire occur. Adequate nonflammable drainage must also be provided in case a tank containing 'white spirit' breaks open. Thus, it is convenient to place electrophoresis tanks on wooden slats, 1½-2 feet off the ground, the area below this being a large zinc-lined trough with drainage. Because of these special

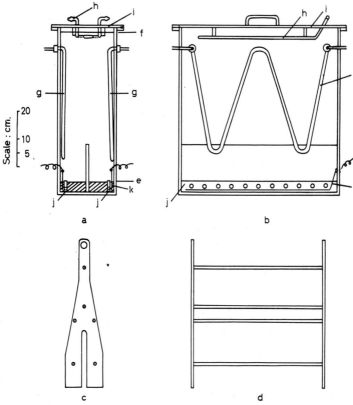

Fig. 3.6. 'Up-and-over' ionophoresis tank and rack (suitable for use with DEAE-papers). (a) front of tank, (b) side of tank, (c) front of rack, (d) side of rack. The tank is made from 1 cm thick Perspex all joints being glued and screwed at 2-inch intervals. It is filled with buffer to the level (e) and with 'white spirit 100' or Varsol to the level (f). Tap water is passed through the stainless steel (g) and glass (h) cooling coils. The latter are attached to the lid (i) which is hinged to the tank. The small partitions (j) are drilled with holes (1.5 cm diam.). Electrodes (k) are of platinum-coated wire. The rack is made of 1 cm thick Perspex (Lucite) and rods of 1 cm diameter.

requirements, it is preferable if electrophoresis tanks are set up in a specially designed electrophoresis room which, in addition to the features already mentioned, should have a large Formica-covered table for 'wetting up' paper and also a sewing machine for stitching papers together. As an additional precaution the floor of such an electrophoresis room should be covered with a welded polyvinylchloride (PVC) sheet. In addition to being inert to the organic solvents and impermeable to water, it is a good insulator.

The most dangerous aspect of high-voltage electrophoresis is undoubtedly the risk of electrocution. However, if access to the tanks is impossible without isolating the power supply, then risk should be negligible. The commonest cause of danger, however, seems to be fire. If the level of coolant in the 'hanging tank' is allowed to drop too low then sparking may occur between the electrode and the surface of the 'white spirit' thus igniting it. Alternatively, if the electrophoresis paper is not completely immersed, it may dry out and sparking may occur. Usually, only a limited fire occurs in the pocket of air between the surface of the coolant and the lid. This danger can be avoided by regularly topping up the tanks, and by avoiding leaving the lids open longer than necessary. Another reason for observing the precaution is to ensure efficient cooling which cannot occur if the coolant is out of contact with the coil. Moreover, uneven runs will result if cooling is not uniform. Some simple instructions displayed in the electrophoresis room of the Laboratory of Molecular Biology, Cambridge, are reproduced here:

1. At the commencement of a run ensure that the cooling tap is turned on, but do not then turn the tap on to its full extent. A steady flow of water is quite sufficient. Also, check that there is enough coolant, i.e. when the lid is down the cooling coil should be in the coolant. If not, fill with coolant provided.
2. If the lid does not close easily do not force it as this will break the cooling coil.
3. At the end of the run turn off the cooling tap, and the appropriate power pack. Close the lid.
4. 'Post-slip' paper is provided for blotting purposes and can be found

on the bottom shelf of the paper rack. Do not use other paper, especially the 'special order' 3 MM paper.
5. Do not discard any solid waste (e.g. Kleenex tissues, remains of broken glass, etc.) into the compartments containing the electrophoresis tanks as this inevitably leads to blockage of the drain underneath.
6. If you are using a 'urea-containing buffer' please clean the glass plate after use.

3.6. Loading, buffers and operating

The resolution obtainable by paper electrophoresis is dependent on several simple precautions. The sample is normally applied as a 2-cm band to the dry paper (except when using a two-dimensional procedure when it is applied as a spot – see ch. 4). When many samples are to be separated, samples should be 2 cm apart, except for very short runs where 1 cm apart is sufficient. It is, thus, convenient to load small volumes of the order of 10 μl. With this low volume, salt concentrations of up to 0.2 M can be tolerated without causing appreciable trailing of bands and a resultant loss of resolution. Larger volumes are much less convenient to apply as this must be done in separate portions, drying down in between applications with a hair dryer. Salt concentrations must also be correspondingly lower. In the radioactive work, samples are normally applied to Whatman No. 540 paper as this paper has a higher wet strength and gives as good a separation as on Whatman No. 1 paper. Whatman No. 3 MM is normally used in the work on non-radioactive RNA because of its higher capacity (see below). This grade of paper is, however, also useful in work with radioactive material particularly when analysing a large number of weakly radioactive samples. Because the No. 3 MM absorbs a given volume of liquid in a smaller area of paper than No. 540 paper, samples may be applied as narrower (1 cm) bands. By reducing the band width from 2 to 1 cm at the point of application the ultimate sensitivity of detection of a weak band is increased 2-fold. Reduction of the band width even further (i.e. by making a point application) is not recommended because it seems easier for the human eye to detect a faint band than a faint point. For the analysis of low amounts of radioactivity the use of DEAE-paper ionophoresis at pH 3.5 is particularly recommended. This is a valuable final analytical step in many procedures (see § 4.4.2.1), because diffusion of oligonucleotides or mononucleotides is smaller than on paper and a large number of samples can be loaded as rather narrow bands. Using this technique, the sensitivity of detection (i.e. the blackness of a band on the radioautograph compared to 'background' blackening) may be up to three times greater than a similar analysis carried out by paper ionophoresis. The origin is marked by pencil points at 1-cm intervals on a line 10 cm from one of the short ends of a 47×57 cm DEAE-paper. Samples are loaded carefully over 1 cm of paper with the point of application accurately on an imaginary line between the two pencil points. Although liquid spreads over a larger area, nucleotides being acidic stick to the DEAE-paper at the point of application. For best results on DEAE-paper, electrophoresis should be performed as slowly as possible, and runs should not be too long. Satisfactory results may obtained on this system using 1 kV for 3 hr. Faster spots diffuse a bit in this time whilst slower moving bands remain very sharp.

The buffers used in electrophoresis should have a significant buffering capacity and should be volatile so that desalting is not required for transfer onto another system. For non-radioactive work a further requirement is that the buffers should not absorb in the ultraviolet regions (240-300 mμ). For radioactive work pyridine-acetate is used for the standard pH 3.5 and pH 6.5 systems, whilst formic acid at high concentration or formic-acetic acid *mixtures* are used for the more acidic runs on DEAE-paper (see ch. 4). The following are the compositions of the buffers for radioactive work; pH 3.5: 0.5% pyridine, 5% acetic acid (v/v); pH 2.1: 2% formic acid, 8% acetic acid (v/v); pH 1.9: 2.5% formic acid, 8.7% acetic acid (v/v); 7% formic acid, pH 1.7 approx.; pH 6.5: 10% pyridine, 0.3% acetic acid (v/v). In the case of the pH 6.5 system, the coolant may be toluene or 'white spirit', otherwise it is 'white spirit'. For non-radioactive work, pyridine buffers are unsuitable as ultraviolet-absorbing impurities remain on the paper after drying off the pyridine-acetate. Usually ammonium formate is used at 0.05-0.1 M for pH between pH 3.0 and 4.5, whilst ammonium acetate is suitable from pH 4-5.5. Using such buffers, very approximately 25 and 75 µg of each nucleotide component of the mixture may be loaded per cm of paper on Whatman 540 and 3 MM paper, respectively, without excessive trailing. Salts in concentration comparable to the buffer will interfere with the separation.

Buffer is applied to the dry-loaded paper by allowing it to chromatograph up evenly as a line from either side of the point of application of the sample, the two fronts meeting evenly at the origin. This procedure requires some practice, but is usually successful if ample time (10-15 min) is allowed. The method has the advantage over 'wet-loading' techniques in that material is concentrated as a narrow band. Buffer is then applied to the rest of the paper, excess buffer blotted off with a cheap grade of filter paper and the paper lifted into one of the electrophoresis tanks. There should be no 'free' buffer remaining on the surface of the paper as this results in uneven fractionation of the sample. This procedure is also adopted with DEAE-paper but requires less care as nucleotides are firmly bound at the origin.

Normally, a voltage gradient of 60 V/cm is used for the separation procedure on paper, although with Whatman No. 540 paper up to 100 V/cm can be used as long as the cooling is adequate (i.e. cooling-coil completely immersed in coolant). Lower voltage gradients of the order of 20 V/cm are used with DEAE-paper (ch. 4). To follow the progress of the electrophoresis, coloured dye markers (appendix 2) applied at the origin may be used. A mixture of 1% xylene cyanol FF, 1% acid fuchsin and 1% methyl orange in water is suitable for mono- and oligonucleotides. When removing the paper from the 'hanging' tank, a glass rod, 60 cm long, is inserted under the paper where it hangs over the rod inside the tank, and is clipped at each end. Excess buffer and 'white spirit' is allowed to drip off before removing the paper, held on the glass rod, from the hood to another well-ventillated fume cupboard for drying. In the case of the 'up-and-over' tank, the rack and paper are lifted out together, the paper being left to dry on the rack.

CHAPTER 4

A two-dimensional ionophoretic fractionation method for labelled oligonucleotides

4.1. Introduction

The work of Holley and others, presented in ch. 2, had shown that progress in the sequence determination of nucleic acids was essentially dependent upon and limited by methods available for the fractionation of the complex mixture of products in enzymatic digests. Although the methods developed in that work were undoubtedly successful, particularly the use of DEAE-cellulose and DEAE-Sephadex ion-exchange chromatography, paper techniques are often preferable from the point of view of speed and ease of operation. The advantages of both techniques may be combined by the use of ion-exchange papers, and Tyndall et al. (1964) have described a two-dimensional technique using chromatography on DEAE-paper at two different pH's. In developing these paper methods further, Sanger et al. (1965) found that good separations were obtained by *high-voltage ionophoresis on DEAE-paper at an acid pH*. It appears that fractionation on this system is due both to ion-exchange and to electrophoretic effects. The positive charges on the paper cause a rapid electro-endosmotic flow of buffer from the cathode to the anode which carries the nucleotides through the paper, thus subjecting them to ion-exchange chromatography. Superimposed on this is the electrophoresis, which is probably an important factor in the fractionation, since considerably better separations are obtained by this technique than by simple chromatography on the DEAE-paper. The ionophoresis opposes the ion-exchange effect, the more acidic components moving faster by ionophoresis but slower by ion exchange.

In conjunction with this method Sanger et al. (1965) have used ionophoresis on *cellulose acetate*. Previous experiments with peptides had indicated that considerably less streaking occurred if cellulose acetate was used instead of paper, and the same was found to be true for nucleotides. A disadvantage of the use of cellulose acetate for many purposes was that only low weights of materials could be fractionated. Nucleotides are normally detected by their absorption of ultraviolet light and this limits the scale on which one can work. In order to lower this scale Sanger et al. (1965) prepared ^{32}P-labelled nucleic acids by biological labelling and studied ribonuclease digests of these, detecting and estimating the oligonucleotides by radioautography and counting techniques. Many nucleic acids can easily be prepared in a radioactive form by biological labelling with ^{32}P-phosphate whether they are from mammalian cells (grown in tissue culture), from bacterial cells, yeasts, or from bacteriophages and details are given in appendix 5.

These two methods, ionophoresis on cellulose acetate and on DEAE-paper were combined in a two-dimensional method which is described in detail in this chapter. The main application of this method is in the determination of the sequence of purified ^{32}P-labelled low molecular weight RNA. The technique is used in the first phase of the sequence determination which is the isolation of the products of a T_1-RNase and a pancreatic RNase digestion of the molecule. A number of such studies have now been reported on purified tRNA molecules and on several 5S RNA molecules. As an example of this application, the methods used to sequence the 5S RNA of *E. coli* will be discussed in full in the next chapter. The method may also be applied to the study of RNA of considerably higher molecular weight than the 5S RNA. For example it may be used to isolate some selected oligonucleotides rather than all of the oligonucleotides from the RNA of bacteriophage R17 (or f2) or from 16S and 23S ribosomal RNA of bacteria. For studying such molecules the method may be described as a 'fingerprinting' procedure and is suitable for comparing species of RNA where differences in sequence rather than the total sequence is required. In this chapter an example of such an application is given: the comparison of the 16S and 23S RNA of *E. coli*. This material, which may readily be prepared, was used by Sanger et al. (1965) to study the potentialities of the method and to determine the position of the different nucleotides on the two-dimensional 'fingerprint'. Digests were prepared by the action of both T_1-ribonuclease and pancreatic RNase.

The 'fingerprinting' technique may also be used to fractionate ^{14}C-labelled oligonucleotides but, because the sensitivity of detection of this isotope is much less by radioautography than for ^{32}P, their sequence cannot usually be determined. However, ^{14}C-labelled material can be used to locate specific ^{32}P-labelled oligonucleotides and such an approach has been used by Fellner and Sanger (1968) for the determination of sequences around methylated bases in 16S and 23S RNA of *E. coli*. Transfer RNA of *E. coli* contains sulphur nucleotides, which may be identified in oligonucleotide fragments of ^{35}S-labelled material using the 'fingerprinting' procedure. A description is, therefore, given in appendix 5 for preparing ^{14}C- and ^{35}S-labelled *E. coli*.

Finally, a more recent application of the method is for the study of RNA labelled in vitro with ^{32}P. Two separate approaches have already yielded useful results and are presented in ch. 9.

4.2. Enzymatic digestion

Digestion has been done with pancreatic and T_1-RNase and also with T_1-RNase and bacterial alkaline phosphatase together. T_1-RNase is more specific than pancreatic ribonuclease and is more generally useful.

The conditions for enzymic digestion are the same for both T_1- and pancreatic ribonucleases, and were such that there were few cyclic nucleotides present which complicate the 'fingerprint'. An enzyme to substrate ratio of between 1 to 10 and 1 to 20 in 0.01 M Tris buffer (*p*H 7.4) containing 0.001 M EDTA (neutralised) was used. Usually carrier transfer RNA was added to the ^{32}P-RNA to make the weight about 20 μg and the digestion was carried out in the tip of a drawn-out capillary tube in a volume of 5 μl for 30 min at 37 °C. The digest could then be applied directly to the cellulose acetate for fractionation in the first dimension. For a combined T_1-RNase and alkaline phosphatase digestion incubation was for 60 min at 37 °C using an enzyme to substrate ratio of 1 to 10 for T_1-RNase and 1 to 5 for bacterial alkaline phosphatase in 0.01 M Tris-chloride, pH 8.0.

4.3. Two-dimensional ionophoretic fractionation procedure
(Sanger et al. 1965)

The digest, containing not more than 0.1 mg nucleotides in a volume of less than 10 μl, is fractionated in the first dimension at pH 3.5 on strips of cellulose acetate by high voltage ionophoresis using the 'up-and-over' tanks described in ch. 3. Cellulose acetate may be obtained in sheets 95 × 25 cm from Oxo, Ltd., although variations in batches do occur and bad batches cause 'tailing' of nucleotides. Pre-cut strips of cellulose acetate are also available from Schleicher and Schüll in 3 × 55 cm long strips. Although the short length (55 cm) of these strips is a disadvantage the resolution obtainable with this source of material is excellent and often better than may be obtained with longer strips of the Oxoid material. 'Cellogel' cellulose acetate sheets from Colab. have also been used successfully by Forget and Weissman (1967). The cellulose acetate strip should be about 3 cm wide and its length will depend on the fractionation required. To obtain a 'fingerprint' with all the smaller nucleotides as in fig. 4.1 the strip should be about 60 cm long. For better fractionation of larger nucleotides as in fig. 4.2 strips of 85 cm can be used. The cellulose acetate was first wetted with the pH 3.5 buffer (0.5% pyridine-5% acetic acid (v/v)) and, to avoid the inclusion of air bubbles, this should be done from one side by flotation. One end of the strip is placed on the surface of the buffer contained in a Petri dish. As it absorbs the buffer the rest of the strip is wetted by passing it slowly across the surface of the liquid. Finally the strip is completely immersed in buffer to ensure it is completely wet. In more recent experiments, a urea-containing pH 3.5 buffer has been used as this helps to avoid streaking observed particularly with some batches of Oxoid and generally improves the resolution of the larger oligonu-

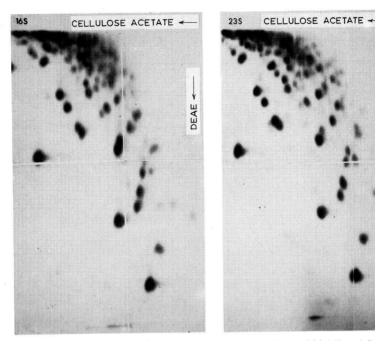

Fig. 4.1. Two-dimensional fractionation of a T_1-RNase digest of (a) 16S and (b) 23S RNA of *E. coli* using ionophoresis on cellulose acetate at *p*H 3.5 in the first dimension; and ionophoresis on DEAE-paper using the *p*H 1.9 mixture in the second dimension running the blue marker approximately half way. For the identification of the spots see fig. 4.5. For a detailed discussion of the differences in these two 'fingerprints' see Sanger et al. (1965).

cleotides. This urea-containing buffer is also used in the buffer compartments of the electrophoresis tank. The buffer was a solution of 5% (v/v) acetic acid and 7 M urea adjusted to *p*H 3.5 with a few drops of pyridine. As the buffering capacity was very low, the *p*H of the solution is not very stable; it is therefore kept at 4 °C, and rejected when its *p*H rises above 3.7. It should be noted, however, that the fractionations shown in the figures in this chapter were obtained with the non-urea-containing buffer and that the urea probably does cause slight loss of resolution of isomeric oligonucleotides in the first dimension. The point of application, about 10 cm from one end of the strip, is blotted

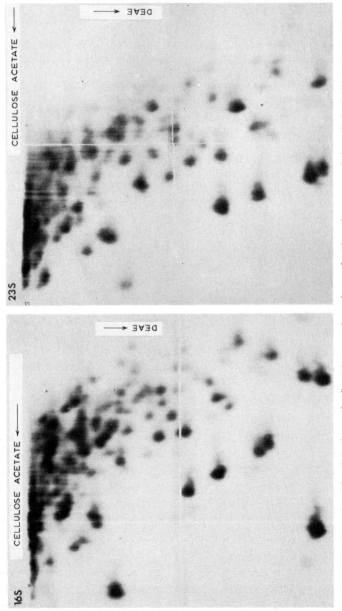

Fig. 4.2. Two-dimensional fractionation as for fig. 4.1 except that a longer fractionation was carried out in both dimensions with a consequent improvement in resolution.

free of excess liquid and the digest is applied as a small spot and allowed to soak in. A dye mixture of Xylene cyanol F. F. (blue), acid fuchsin (red) and methyl orange (yellow) (all from G.T. Gurr, Ltd.) is applied on each side of the digest. Care has to be taken to avoid the strip drying out while the sample is applied; this is done by having the ends dipping in the buffer or covered with wet tissue paper. Excess buffer is then removed from the strip by blotting and it is rapidly dipped in 'white spirit' (Varsol) to prevent evaporation of the buffer while the strip is put into the ionophoresis tank. The origin is at one end near the negative electrode vessel. Ionophoresis is carried out at between 50 and 100 V/cm. For the shorter strips where the distance between the two electrode vessels is 50 cm electrophoresis is continued until the yellow marker has almost reached the anode buffer (about 1 hr at 100 V/cm). It was found that in T_1-RNase digests the majority of the nucleotides migrate slower than the yellow and faster than the blue marker. In order to achieve better resolution in the area between these two markers, a longer run is possible (2 hr at 100 V/cm) using 85 cm strips and only that part between the blue and pink markers (about 40 cm) is used for the second dimension. The shorter length of cellulose acetate conveniently fits over the lower part of the ionophoresis rack whilst the larger strips fit over the upper section (fig. 3.6).

Fractionation in the second dimension is carried out on sheets of DEAE-paper (Whatman DE 81). Pre-cut sheets, 57×47 cm, are available and were used in the fractionation shown in figs. 4.1 and 4.2. However, longer sheets, 85×47 cm cut from a 100 metre roll of DE 81, allow longer runs with an improvement in resolution, especially of the slower-moving oligonucleotides and are, therefore, preferable. The DEAE-paper is rather fragile and difficult to handle when wet; cellulose acetate on the other hand is easy to handle when wet but brittle when dry. Attempts to sew the cellulose acetate strip onto the DEAE-paper were unsuccessful. A good transfer of the material from the cellulose acetate can, however, be obtained by the following blotting procedure. The sheet of the DEAE-paper is laid on a glass plate. The strip of cellulose acetate on which the nucleotides have been fractionated is removed from the tank and hung up so that the excess 'white spirit' drips off. Before the buffer dries the wet strip is laid on the DEAE-cellulose sheet about 10 cm from one of the short sides. A pad of four strips (46×2 cm) of Whatman No. 3 MM paper that has been soaked in water is then put on top of the cellulose acetate strip and a glass plate placed on top to press the strips together evenly. Water from the paper pad passes through the cellulose acetate and into the DEAE-paper carrying the nucleotides with it. Because they are acidic, nucleotides are held to the DEAE-paper by ion exchange and remain in the position in which they are first washed on. One thus obtains a good transfer of the fractionated nucleotides without any smearing of the spots as usually occurs with blotting techniques. By this means, a strip of the DEAE-paper about 6 cm wide is wetted. To ensure a more complete transfer, more water can be added from a pipette to the paper pad while still in position. In a control experiment it was shown that 90-95% of a T_1-RNase digest of RNA could be transferred from cellulose acetate to DEAE-paper by this technique. After removal of the pad and the cellulose acetate strip (after 5-10 min), the positions of coloured spots are marked on the DEAE-paper and it is hung up to dry in a current of warm air.

When the urea-containing buffer is used for the first dimension, urea is transferred in this blotting procedure and has to be removed before fractionation in the second dimension is attempted. This is done by rinsing the wetted area on the DEAE-paper with 95% ethanol in a trough for about 1 min. Failure to remove urea results in a significant loss of resolution of the faster-moving nucleotides in the second dimension.

After drying, the DEAE-paper is wetted with either a pH 1.9 mixture (2.5% formic acid, 8.7% acetic acid (v/v)) or by 7% formic acid. The pH 1.9 mixture was used for studying pancreatic ribonuclease digests whilst both systems have been used for T_1-RNase digests. Due to the fragility of the DEAE-paper the ionophoresis had to be carried out in an apparatus in which the paper was supported on a rack that could be lifted in and out of the tank (see ch. 3).

Wetting of the paper is started on either side of the line on which the nucleotides are applied by squirting liquid from a soft plastic bottle and the fronts of the solution are allowed to meet along this line. After wetting the rest of the cathode end of the paper (i.e. one-third of the total length of paper), it is put on the rack for running in the ionophoresis tank. The rest of the paper is wetted on the rack which is then lowered slowly into the tank with the applied nucleotides near the cathode compartment. Both solvent systems have a high conductivity and there is considerable heating if very high voltages are used. To obtain a 'fingerprint' as in fig. 4.1 using the pH 1.9 mixture, about 30 V/cm were used until the blue marker was almost at the top of the rack (about 4 hr) on a short DEAE-paper. For better separations as in fig. 4.2 using the pH 1.9 mixture the ionophoresis was carried out overnight at 30 V/cm, by which time the blue marker had usually run off the end of the paper. For separations using 7% formic acid lower voltages (up to 20 V/cm) have to be used and runs take up to 24 hr. The DEAE-paper is finally dried in air while still on the rack. To prepare radioautographs the paper is marked with radioactive ink (red ink containing ^{35}S-sulphate) and put in a folder with one or two sheets of X-ray film. The folders are covered on one side with a thin sheet (0.5 mm) of lead and stacked in a light-proof cabinet. Where more than 1 μC of ^{32}P-RNA had been used the film could be developed after 12 hr, and there would be sufficient material to study the structure of the nucleotides. If it was only desired to obtain a 'fingerprint', less material could be used (about 0.01 μC) and the film developed after 1-2 weeks.

The conditions for the DEAE-paper ionophoresis are rather acidic and some depurination of the nucleotides might be expected. That this cannot be very extensive is apparent from the fact that all expected nucleotides are found in the digests. It would seem that the effective pH on an ion-exchange material is not necessarily that of the buffer with which it is washed. Occasionally after a long run at a high-voltage gradient the tank may become warm and streaking may be observed behind some of the spots which may be due to depurination.

4.4. Structure of nucleotides

The positions of nucleotides on the DEAE-paper is determined from

Subject index p. 261

the radioautograph using the marks from the radioactive ink to line up the film with the paper. The spots are cut out and eluted with alkaline triethylamine carbonate prepared as follows: CO_2 is passed into a mixture of 70 ml water and 30 ml redistilled triethylamine until a single phase of triethylamine carbonate formes. More triethylamine is then added with shaking to pH 10. In order to minimise losses due to the very small amounts of nucleotides present, non-radioactive nucleotides are added as carrier to the eluting medium. These were prepared as follows: 1.0 g yeast RNA (appendix 2) was treated with 25 ml 1.0 N KOH at 37 °C for 20 min. The KOH was then neutralised with concentrated HCl and dialysed for about 4 hr against distilled water. The contents of the dialysis membrane were made up to 50 ml and thus contained 20 mg nucleotides/ml. 2 ml of this was added to each 100 ml of the eluting mixture.

Elution is carried out as described by Sanger and Tuppy (1951) except that thin-walled capillary tubes are used (prepared by drawing out open-ended melting point tubes). Elution is very rapid by this method and within 20 min about 50 μl is collected containing about 20 μg of carrier nucleotides. The eluted material is put onto polythene sheets and either allowed to evaporate overnight, or taken to dryness in a desiccator containing a beaker of H_2SO_4, and a Petri dish with NaOH pellets in a partial vacuum (if a good vacuum is used the triethylamine bubbles badly). In order to ensure complete removal of the triethylamine carbonate, water is added to the spots and they are taken to dryness several times. The spot is then divided for analysis by one or other of the procedures below. Further procedures, such as U_2-RNase, T_2-RNase, acid RNase and pancreatic RNase digestion of carbodiimide-blocked nucleotides are described in the succeeding chapters.

4.4.1. Composition

In order to identify the mononucleotides present in an oligonucleotide it is digested in drawn-out melting point tubes with about 10 μl 0.2 N NaOH at 37 °C for 16 hr. The solution is allowed to enter the tube by capillarity. The liquid is drawn into the middle of the tube by connecting one end to a piece of polythene tubing and then creating a negative pressure. This is achieved by drawing the thumb nail across the polythene tube held firmly between thumb nail and forefinger. Both ends of the tube are then sealed off in a flame. After incubation the digest is applied to Whatman No. 540 or 3 MM paper for ionophoresis at pH 3.5 and 60 V/cm for 1 hr. Digests were usually applied over 2 cm of the paper with 2 cm gaps between them (but see § 3.6). The four mononucleotides are well separated on this system and the composition of each digest can be determined from the radioautograph. With nucleotides from T_1-RNase digests, G was present as a single spot, but with those from pancreatic ribonuclease digests it formed two spots, which were the 2'- and 3'-phosphates. C was also split into its 2'- and 3'-phosphates in good runs. With simple nucleotides it is usually possible to deduce how many residues of each mononucleotide are present by visual estimation of the intensity of the bands on the radioautograph; however, this is best checked by direct estimation in a scintillation counter. Equal areas of the various bands are cut out from the paper using the radioautograph as a guide. The radioactivity of each area is counted in vials with toluene containing 0.4% BBOT (Ciba). If the spots contain less than about 100 counts/min the results are not very reliable, but above this about 10% accuracy can be achieved.

4.4.2. *Digestion with enzymes*

To determine the sequence of residues in oligonucleotides they are further degraded with the nucleases described below and the products identified. The most useful enzyme for the investigation of nucleotides from T_1-RNase digests is pancreatic ribonuclease digests whilst the most useful enzyme for studying products of pancreatic RNase digests is T_1-RNase. In both cases the products that can be obtained are the same and consist of the mononucleotides G, C and U and oligonucleotides with various numbers of A residues terminated at their 3'-end by G, C or U (see fig. 4.3).

4.4.2.1. Pancreatic ribonuclease. Samples of the oligonucleotides are dried down on polythene and treated with 5-10 μl of 0.001 M EDTA, 0.01 M Tris buffer (*p*H 7.4) containing 0.1 mg pancreatic ribonuclease/ ml and transferred to capillary tubes which are then sealed at one end only and incubated at 37 °C for 30 min. Samples are then treated with 0.1 N HCl to break down any cyclic phosphates as follows: 2 μl 0.5 N HCl is put on a polythene sheet, the closed end of the capillary is broken and the digest squirted out and mixed with the HCl, and then sucked back into the capillary which is resealed and incubated at 37 °C for 1 hr. The digest is then transferred to Whatman No. 540 paper and ionophoresed at *p*H 3.5 at 60 V/cm until the pink marker (see appendix 2) is near the end of the paper. The products may be clearly identified from their position (fig. 4.3) except for AG, AAG and AAAG which are only just resolved on the ionophoresis. If there is any doubt, the band in the AG position is eluted and subjected to ionophoresis on DEAE-paper using either the *p*H 1.9 mixture or 7% formic acid. The nucleotides are then clearly separated according to their size. The acid (HCl) treatment may be omitted in which case C⟩ and U⟩ are present in addition to C and U (fig. 4.3). More recently ionophoresis at *p*H 3.5 on DEAE-paper has been used for separating products of pancreatic RNase digests of T_1-ribonucleotides and the mobilities on this system are shown in fig. 4.4. This system is a compromise between the *p*H 3.5 paper ionophoresis system and the DEAE-paper, *p*H 1.9 system. A particular advantage of using DEAE-paper, rather than paper, is that nucleotides run as rather more compact bands on the ion-exchange paper and hence the sensitivity of detection of rather weakly radioactive samples in higher than with paper (see § 3.6).

4.4.2.2. T_1-RNase. Oligonucleotides from pancreatic ribonuclease digests are digested with T_1-RNase using 0.001 M EDTA, 0.01 M Tris buffer (*p*H 7.4) and 0.1 mg/ml enzyme for 30 min at 37 °C and fractionated as described above in § 4.4.2.1.

4.4.2.3. Micrococcal nuclease. Digestion is carried out in 0.05 M borate buffer (*p*H 8.8) containing 0.01 M $CaCl_2$, with an enzyme concentration of 0.1 mg/ml. Incubation is for 2 hr at 37 °C and the prod-

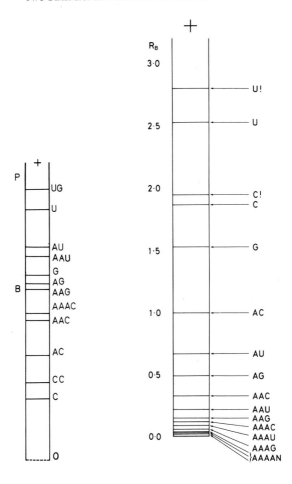

Fig. 4.3. Positions of P-RNase products of T_1-oligonucleotides (or T_1-RNase products of pancreatic oligonucleotides) on ionophoresis on Whatman No. 540 paper at pH 3.5. B and P are the blue (xylene cyanol FF) and pink (acid fuchsin) markers. O is the origin. CC and UG are derived from micrococcal nuclease digests of CCUG.

Fig. 4.4. Position of products of T_1-RNase digests of pancreatic oligonucleotides (or vice versa) on ionophoresis on *DEAE-paper at pH 3.5*. R_B is mobility relative to blue (xylene cyanol FF) marker.

ucts are fractionated at pH 3.5 on paper or DEAE-paper. This method had only a limited use as the specificity of this enzyme is not clearly defined (Reddi 1959).

4.4.2.4. Venom phosphodiesterase. Treatment of oligonucleotides *lacking a 3'-terminal phosphate group* (from T_1-RNase and phosphatase digests) is carried out with 5-10 μl of the following solution: 0.01 M magnesium acetate, 0.05 M Tris buffer (pH 9.0) and 0.1 mg venom phosphodiesterase/ml. After incubation for 1 hr at 37 °C, the 5'-mononucleotides are separated by paper ionophoresis at pH 3.5. The mononucleoside 5'-phosphates move slightly slower than the corresponding mononucleoside 2'(3')-phosphates. Venom phosphodiesterase hydrolysis is the standard method of determining the 5'-terminal residue of dephosphorylated oligonucleotides obtained from a combined T_1-ribonuclease and bacterial alkaline phosphatase digestion. *The 5'-terminal residue is released from the oligonucleotide as a nucleoside and is identified as that residue which is missing from the venom phosphodiesterase digest but present in the alkaline digest.*

The 5'-terminal residue of oligonucleotides *with a 3'-terminal phosphate* may be found by digestion with venom phosphodiesterase under more violent conditions (Holley et al. 1965b). Digestion in this case is with 0.2 mg/ml of enzyme in 0.05 M Tris-chloride 0.01 M magnesium chloride pH 8.9 for 2 hr. at 37 °C. The 5'-end group is released as a nucleoside (and is therefore not seen) whilst the 3'-end is released as a mononucleoside 3', 5'-diphosphate. The four diphosphates have mobilities relative to uridine 5'-phosphate at pH 3.5 on Whatman No. 540 paper of pCp, 1.05; pAp, 1.1; pGp, 1.3; pUp, 1.8. One difficulty of this procedure is that it is apparently particularly susceptible to 'overdigestion' – a cleavage between -PypA- bonds liberating -Pyp and A- (see § 5.2.1) – so that the results must be interpreted with caution.

An alternative procedure for the 5'-end-group determination of oligonucleotides with a 3'-terminal phosphate group, is to remove this phosphate by treatment with bacterial alkaline phosphatase (1 part of enzyme to 10 parts of substrate for 0.5 hr at 37 °C in 0.01 M

Tris-chloride, pH 9.0). The phosphatase is then inactivated by adding 1/10 volume of 0.01 M EDTA and heating for 3 min at 90 °C, after which snake venom phosphodiesterase and magnesium chloride are added to a final concentration of 0.1 mg/ml and 0.01 M, respectively, and digestion is continued for a further 1 hr at 37 °C (Min-Jou and Fiers 1969).

4.4.2.5. Partial digestion with spleen phosphodiesterase. The conditions of digestion depended on whether the oligonucleotide had been isolated from a T_1-RNase or pancreatic RNase digest. For nucleotides from a T_1-RNase digest all ending in guanosine 3-phosphate the following conditions were found to be the most generally useful and have been used in most preliminary experiments. The nucleotide is treated with 5-10 μl of the following mixture in a capillary tube for 30 min at 37 °C: 0.3 M ammonium succinate (pH 6.5) or 0.1 M ammonium acetate, pH 5.7, 0.002 M EDTA, 0.05% Tween 80, 0.2 mg enzyme/ml (appendix 2). Oligonucleotides containing more than one C residue are relatively resistant to digestion, which is then usually carried out for 60 min. Nucleotides from pancreatic RNase digests are digested more rapidly, and for satisfactory results an enzyme concentration of 0.1 mg/ml is used for 30 min. Because of variations in activity of different batches of enzyme and of the different susceptibility of oligonucleotides to degradation, various conditions should be tried (e.g. different times of digestion, or different enzymic concentrations).

Min-Jou and Fiers (1969) have found in their work on large polypurine sequences that an improved spectrum of degradation products may be obtained if the sample is heated briefly (3 min at 90 °C) in the buffer (0.3 M ammonium succinate-0.002 M EDTA-0.05% Tween 80, pH 6.5) before adding the enzyme. This procedure is designed to break up aggregates which are presumably less sensitive to degradation with the enzyme.

After digestion the material is applied directly to DEAE-paper with an untreated control sample applied next to it with no space between. A sample of an unfractionated T_1-RNase digest of any convenient source of RNA is also applied to each paper to serve as a

Subject index p. 261

marker for the position of oligonucleotides. Coloured markers are applied to allow the course of the ionophoresis to be followed visually. After wetting with the pH 1.9 mixture, ionophoresis is carried out usually on short sheets as described in § 4.3 above, until the blue marker (see appendix 2) has reached the top of the rack (about 4 hr at 30 V/cm).

4.5. *Application of the two-dimensional method for finger-printing 16S and 23S rRNA*

The potentialities of the fractionation system (described in § 4.3) are illustrated in figs. 4.1 and 4.2. The sequences of the spots in these fingerprints (Sanger et al. 1965) is deduced by application of the methods in § 4.4. The logical deduction of the sequence of all possible di-, tri- and tetranucleotides, present in complete T_1- and pancreatic ribonuclease digests, has already been presented (table 2.2 and § 2.6) and will not be discussed here further. The reader should note carefully the use of the word *complete* T_1-RNase digestion to indicate that all the products (termed end products of T_1-RNase digestion) have only one G residue –at their 3'-phosphate end. This is in contrast to *partial* digestion products, derived from partial T_1-RNase digests which, in addition to a 3'-terminal G have one or more *internal* G residues. The following discussion refers only to end products of T_1- and pancreatic ribonuclease digestion. For studying pentanucleotides isolated on the two-dimensional system from rRNA, and for confirming the structure of the tetranucleotides, it is useful to use the method of partial digestion with spleen phosphodiesterase, a method developed by Sanger et al. (1965) especially for the study of radioactive oligonucleotides. For sequences longer than pentanucleotides encountered in complete T_1- and pancreatic ribonuclease digests, other methods are preferable (§ 5.2.2). The use of spleen phosphodiesterase is presented below (§ 4.6) after a discussion of the general features of the two-dimensional fractionation system.

4.5.1. T_1-RNase digests

Figs. 4.1 and 4.2 show short and long runs of T_1-RNase digests of the separated ribosomal RNA's using the pH 1.9 mixture in the second dimension. The di-, and trinucleotides are all readily separated by this technique as can be seen from the 'maps' shown in fig. 4.5. Of the 27 possible tetranucleotides (table 2.2) 10 are present as single pure spots. Very slight overlapping usually occurs between CUCG (23) and CCUG (24) and between AUAG (32) and AAUG (33). The following pairs occur in single spots, where there is slight evidence of separation; AUCG (29) and ACUG (30), ACAG (19) and AACG (20). No separation has been observed between the pairs UCAG (25) and UACG (26), and UUAG and UAUG (37). The three isomers having the composition $(CU_2)G$ (34-36) are in a large spot which shows partial resolution of all three. The composition of the pentanucleotide spots has been determined in many cases, but the sequences have not been studied in detail and it seems probable that most spots are mixtures of isomers. There are 81 possible pentanucleotides, and for some compositions – e.g. $(C_2AU)G$ – there are twelve possible isomers. There is a considerable spread of isomers as is indicated by the difference in the patterns in the pentanucleotide area obtained from the 16S and 23S components. The areas occupied by the different isomers are indicated in fig. 4.5b with dashed lines.

The position of a nucleotide on the 'fingerprint' is determined largely by its composition. Fig 4.6 shows the relationship between the composition of a nucleotide and its position, using the pH 1.9 mixture in the second dimension. It was prepared as follows. The 'centre of gravity' or mid-point of all oligonucleotides with the same composition (i.e. isomers) was marked on tracing paper overlaying the fingerprint. These isomers were always close to one another. Lines were then drawn to connect the resultant points with other points representing the position of oligonucleotides with increasing numbers of C residues. Thus, for example, a line was drawn connecting UG, (CU)G (mid-point of two isomers), $(C_2U)G$ (mid-point of three isomers), $(C_3U)G$ (mid-point of four isomers), etc. Similarly, lines were drawn connecting nucleotides differing only in numbers of A residues (e.g. UG, (AU)G,

Subject index p. 261

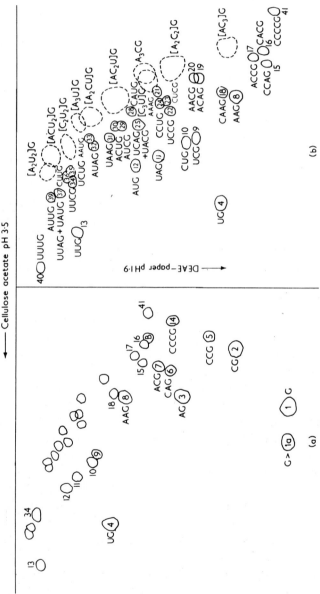

Fig. 4.5. Diagram showing the position of nucleotides from a T_1-RNase digest of rRNA on the two-dimensional system using the pH 1.9 mixture in the second dimension. (a) A fractionation that has been run for a short time on the DEAE-paper. (b) One that has been run for a longer time. B marks the position of the blue marker.

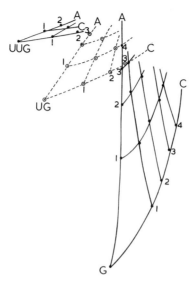

Fig. 4.6. Diagram illustrating the relationship between the composition of a nucleotide and its position on the two-dimensional system using ionophoresis at pH 1.9 for the DEAE-paper dimension. For explanation see text.

$(A_2U)G$), etc. In some cases, where there was only one possible isomer, lines could be drawn connecting spots on the fingerprint directly – for example between G, CG, CCG, CCCG, etc., and between G, AG, AAG, AAAG, etc. From this analysis (fig. 4.6) it emerged that the mononucleotide that has the greatest effect on mobility in the DEAE-paper dimension is U so that the 'fingerprint' may be regarded as being composed of three sections representing nucleotides with two, one or no U residues respectively. The lines joining the spots form three graticules corresponding to the different sections and one axis on each graticule represents the number of A residues, whereas the other gives the number of C residues. *In this way the probable composition of a nucleotide may be determined from its position on the map.* In the pentanucleotide area there is considerable overlapping between the different graticules. Thus, for instance, $(A_3C)G$ is in the same area as $(C_3U)G$. This overlapping may be almost completely avoided by carrying out the fractionation on DEAE-paper using 7% formic acid instead of the pH 1.9 mixture. This increases the separation of the sections containing different numbers of U residues, and a distribution as shown in fig. 4.7a is obtained. The resolution of isomers of the faster-moving spots is probably less efficient on this system, although it is definitely better for the slow-moving ones. *This 7% formic acid system has been used extensively in other applications of the method, and has largely superseded the pH 1.9 system for studying T_1-RNase digests, principally because of the better resolution of the larger oligonucleotides.*

Oligonucleotides containing two or more U residues move slowly on DEAE-paper even when 7% formic acid is used, and fractionation of these is relatively poor. However, if a combined T_1-RNase and bacterial alkaline phosphatase digestion is done before fractionation, the dephosphorylated products (lacking a 3'-terminal phosphate) move much faster than the corresponding phosphorylated nucleotides (fig. 4.7b). Thus dephosphorylated nucleotides with two U residues run in the same region as phosphorylated nucleotides with one U residue. Consequently, oligonucleotides with two or more U residues are better resolved in the dephosphorylated form. For example, U-U-A-G and U-A-U-G separate whereas U-U-A-G- and U-A-U-G- do not (see spot 37 of fig. 4.5). Fig. 4.7b shows that dephosphorylated nucleotides with no U residues run more slowly on both dimensions than normal and form a 'tail' to the overall pattern of spots. Presumably the effect on the DEAE-dimension is a result of the ionophoretic properties of the system overriding the ion-exchange properties. For example, C-C-G moves faster than C-G. This should be compared with the same nucleotides having a 3'-terminal phosphate where the reverse is true, C-G- being faster than C-C-G-. In general there is a separation of isomers where those isomers which have A as the 5'-terminal residue move slower in the DEAE-dimension than those having other 5'-terminal residues.

The advantage of fractionating oligonucleotides lacking a 3'-terminal phosphate is well illustrated by the results of Fellner et al. (1970) who have isolated *all* the end products of T_1-ribonuclease digestion of 16S RNA by this method.

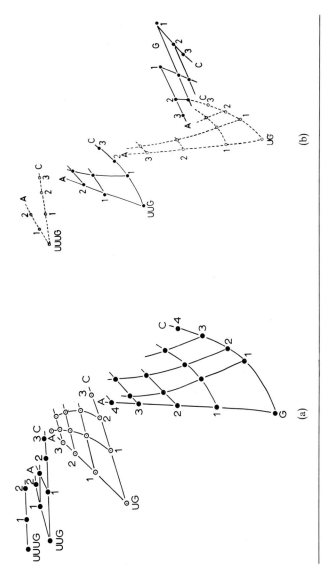

Fig. 4.7. Diagram illustrating the relationship between the composition of a nucleotide and its position on the two-dimensional system using 7% formic acid for the DEAE-paper dimension; (a) is for nucleotides in T_1-RNase digests and (b) for nucleotides (lacking a 3'-terminal phosphate residue) in combined T_1-RNase and alkaline phosphatase digests.

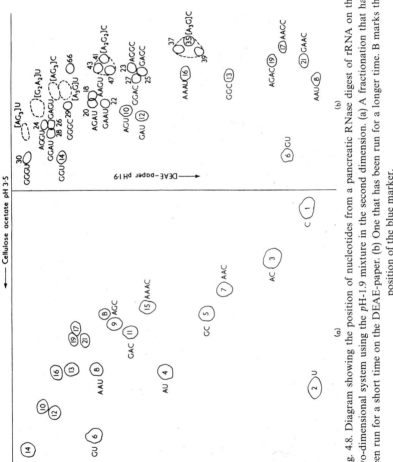

Fig. 4.8. Diagram showing the position of nucleotides from a pancreatic RNase digest of rRNA on the two-dimensional system using the *p*H-1.9 mixture in the second dimension. (a) A fractionation that has been run for a short time on the DEAE-paper. (b) One that has been run for a longer time. B marks the position of the blue marker.

4.5.2. Pancreatic RNase digests

Fig. 4.8 shows a map of the nucleotides on both short and long runs on the DEAE-dimension of pancreatic ribonuclease digests of ribosomal RNA. The simplicity in the composition of the nucleotides and the smaller number of larger ones made a study of these digests somewhat simpler than with T_1-RNase. All possible mono- di- and trinucleotides are separated (fig. 4.8) whilst of the sixteen possible tetranucleotides ten have been identified as single pure spots. The remaining six tetranucleotides show partial separation. Thus, the following pairs of isomers show partial separation but there is usually some overlapping: AAGU (18) and AGAU (20), GAGC (25) and GGAC (27), GAGU (26) and GGAU (28).

It was always possible to identify the 5'-terminal residues from the positions of the nucleotides on the 'fingerprint', since isomers with A end groups move slower on the DEAE-cellulose than those with G end groups. For example, AGU, spot 10 is slower than GAU, spot 12 of fig. 4.8b. The 7% formic acid system may also be used in the second dimension for pancreatic RNase digests and as in the T_1-RNase digests there is an improvement in resolution of the larger oligonucleotides at the expense of a decreased resolution of the smaller oligonucleotides. The best compromise for an analysis of high molecular-weight RNA is probably to isolate the faster-moving oligonucleotides on the pH 1.9 system, whilst analysing the slower ones by means of 7% formic acid. For a small molecular weight RNA, either system may be used (ch. 5).

4.6. Results of partial spleen phosphodiesterase digestion of oligonucleotides

It has been found that on the DEAE-paper-pH 1.9 system used for the two-dimensional procedure any phosphorylated nucleotide will move faster than a corresponding nucleotide having the same structure, but with one extra residue added to its 5'-terminal end. Consequently, the various degradation products produced by an exonuclease on an oligonucleotide will be arranged in order of their size on fractionation in this system, the larger ones moving slower than the smaller ones. The distance between any two products which differ by only one residue will depend on the nature of that residue. *Thus, from these distances one may deduce the nature of the various residues, and hence it is frequently possible to determine a complete sequence by a single degradation and ionophoretic fractionation.* This method has been used after partial digestion with spleen phosphodiesterase which splits off 3'-mononucleotides sequentially from the 5'-terminus of a phosphorylated oligonucleotide (Razzell and Khorana 1961). It has been applied to oligonucleotides in complete digests from both T_1- and pancreatic RNase digests of RNA. A somewhat similar method using venom phosphodiesterase has been described by Holley et al. (1964) and details of this procedure applied to radioactive 5S RNA appear in ch. 5.

4.6.1. Partial spleen phosphodiesterase digestion of oligonucleotides derived from complete T_1-RNase digestion

Some of the results of spleen enzyme treatment of T_1-RNase digests are illustrated in fig. 4.9a. The numbers refer to the spots on fig. 4.5. If we consider the nucleotide ABCG where A, B and C are unknown residues, its breakdown products will be BCG and CG. Suppose the distance of ABCG from the origin of the DEAE-paper is y and the distance between it and its first degradation product (BCG) is x, we can then define a value M for ABCG as x/y. This value depends on the nature of the 5-terminal residue A. For all the oligonucleotides studied from the T_1-RNase digests it has been found that the M values lie within the limits shown in table 4.1 and there is no overlapping between the

TABLE 4.1

Values of M on DEAE-paper ionophoresis (pH 1.9) for oligonucleotides in complete T_1- and pancreatic RNase digests.

5'-terminus	Range of M values
C	0.05–0.3
A	0.4 –1.1
U	1.5 –2.5
G	1.2 –3.5

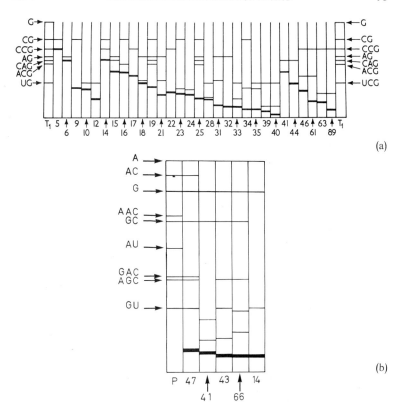

Fig. 4.9. Positions of products of partial spleen phosphodiesterase digestion fractionated by ionophoresis on DEAE-paper using the pH 1.9. mixture of (a) some T_1-oligonucleotides derived from the experiment illustrated in fig. 4.5, and (b) some pancreatic oligonucleotides derived from the experiment illustrated in fig. 4.8. The thick line represents the position of the untreated nucleotide which was run as a control. T_1 is a marker T_1- and P a marker pancreatic RNase digest of rRNA to show the positions of some of the di-, and trinucleotides. (See appendix 6 for sequences of some products.)

values for the three different 5'-terminal residues, C, A and U. (It should be noted that G can never be 5'-terminal in oligonucleotides derived from complete T_1-RNase digests and that the figures for a 5'-terminal G quoted in table 4.1 derive from studies described below (§ 4.6.2) on products of complete pancreatic RNase digestion.) Thus, the various residues may be identified directly from the M values. The dinucleotides and many of the trinucleotides can be identified from their position on the DEAE-paper by comparison with the untreated T_1-RNase digest that is run as a marker.

As examples we may consider the results with the three isomers $(C_2A)G$, numbers 15, 16 and 17 (fig. 4.9a). The M value for the first degradation step for 16 is 0.18. This falls within the limits for C (table 4.1) which is therefore the 5'-terminal residue. The M value for the second step is 0.58 establishing it as an A residue. The product at this stage is rapidly moving and can only be the dinucleotide CG, thus establishing the structure CACG. For No. 15 the M values for the first two steps are 0.22 and 0.07, respectively, and the dinucleotide moves at the rate of AG, thus giving the sequence CCAG. Nucleotide No. 17 gives only one visible degradation product under the conditions used. This is related to the unchanged material by an M value of 0.64 and occupies a unique position which identifies it as the trinucleotide CCG, thus confirming the structure ACCG.

For tetramers and some pentamers it is possible to deduce the complete sequence by this method. Difficulties may arise if the oligonucleotides studied are mixtures. Thus the results (fig. 4.9a) with No. 19 (fig. 4.5) could be interpreted wrongly. The M values of the various degradation steps for spot 19 are 0.64, 0.08, 0.06 and 0.29. The final product is rapidly moving and in a position corresponding to CG in the marker digest. From the M values of table 4.1 one would deduce that the sequence was ACCACG. This result, however, is incompatible with the observed position of spot 19, relative to other oligonucleotides, on the fingerprint. The composition of spot 19 predicted, as discussed in § 4.5.1, from its position, is $(CA_2)G$. A closer examination of the results for spot 19 in fig. 4.9a shows in fact that a band is present in the position of AG in addition to the one in the position of CG. This can only mean that spot 19 is a mixture of at least two isomers. The M values can now be reinterpreted as follows. The first degradation corresponding to an M value of 0.64 indicates an A-terminal residue. The product of this degradation is in the position corresponding to ACG in the marker digest. Further degradation gives CG. Thus AACG is present. The other band derived from spot 19 is also in a unique position corresponding to CAG in the marker digest which degrades further to give AG (already noted). The sequence of this isomer is thus ACAG. It will be noticed that the third isomer with the composition $(CA_2)G$, which is CAAG, is present in the nearby spot 18 and is clearly absent from spot 19 as there is no evidence of any C-terminal residue. Spot 25 is another example of a mixture of isomers. Where it is suspected that a mixture is present it is often useful to write down the possible isomers and then to see which can be excluded from the results. Spot 25, from its position on the fingerprint (see figs. 4.5 and 4.6) would have the composition (CAU)G. It could therefore be one of six isomers: (1) CAUG; (2) CUAG; (3) ACUG; (4) AUCG; (5) UACG; (6) UCAG. The M value for the first degradation (fig. 4.9a) is 1.75 indicating a 5'-terminal U residue. This observation, together with the absence of any products corresponding to 5'-terminal C or A residues, suggests that sequence (5) or (6) is present. In fact, both are present because bands corresponding to ACG and CG, as well as to CAG and AG are among the degradation products.

Sometimes additional bands were present in the nucleotide control which was run side by side with the sample treated with spleen phosphodiesterase. These arose, presumably, during elution or storage of the oligonucleotide. Such bands, especially if present in good yield, are confusing and can lead to wrong interpretations. Their presence emphasises the need to run such a control. Bands common to the experimental digest and the control are ignored in interpreting the results in the experimental digest, although occasionally they can confuse the issue if they happen to coincide with a 'correct' degradation product. Another difficulty is that occasionally 'incorrect' specificity of cleavage by spleen phosphodiesterase is observed. For example, CCUG gives CCU in low yields in addition to CUG.

One limitation of the method of partial digestion with spleen phosphodiesterase lies in the different rates at which the three different residues are released by the enzyme, U residues being split off more rapidly than A and A more rapidly than C. Thus, not all the possible degradation products of a nucleotide are necessarily present in a

Subject index p. 261

particular digest and this may lead to a misinterpretation in the case of larger nucleotides. For this reason it is advisable not to rely completely on this method but to use it in conjunction with other information, such as the results of alkaline hydrolysis (§ 4.4.1) or pancreatic RNase digestion (§ 4.4.2.1), or determination of the 5'-terminal residue (§ 4.4.2.4). For example in the degradation of spot 17 (fig. 4.9a) the first step had an M value of 0.63 and therefore indicates an A residue. The product of this degradation is in the unique position of the marker CCG, although no CG is observed. This merely means that the rate of cleavage of CCG to CG is so slow under the conditions employed, that the digestion effectively stops at the trinucleotide, CCG. An exactly comparable result was obtained in the case of spot 22, although in this case the M value for the first cleavage was 1.8 and indicated that the sequence was UCCG.

The results in spot 24 are rather difficult to interpret. Only one degradation product was observed, the M value (0.2) indicating a 5'-terminal C residue. As the composition of spot 24 is $(C_2U)G$ (derived by alkaline hydrolysis, or from a knowledge of its position on the fingerprint), this degradation product must be either CUG or UCG. From the results of spot 24 it is not possible to distinguish between these two possibilities. However, spot 23 is clearly CUCG and would therefore suggest by difference that spot 24 was the other isomer CCUG with a 5'-terminal C residue. The reason for the absence of UG among the digestion products must therefore, again, reflect the slow rate of hydrolysis of C residues under the conditions used for partial digestion. In fact, it is possible to distinguish the isomeric trinucleotides UCG and CUG, the former being slightly faster than the latter (see spots 9 and 10 of Fig. 4.5), so that a comparison of the first degradation product in spot 24 with the first product of spot 23 would again support the result that 24 is CCUG.

Spot 32 (fig. 4.9a) is another example which could lead to difficulties. The first degradation has an M value of 4.4 which is outside the possible range for any of the three (C, A and U) 5'-terminal residues. In this case one must assume that an intermediate degradation product between the unchanged material and the first observable degradation product (which is in the position of AG), is missing. As the composition of spot 32 is $(A_2U)G$, derived by experiment or from a knowledge of its position, the sequence must be either AUAG or UAAG. These possibilities cannot be distinguished in this experiment except by reference to results with other spots. For example, spot 31, from the M values, is clearly UAAG and would, by difference, suggest that spot 32 were AUAG. This postulate could most easily be confirmed from a knowledge of the products of pancreatic RNase digestion of spot 32 (§ 4.4.2.1) which does, in fact, give the correct products – AU and AG. The absence of the intermediate UAG in the digestion of AUAG is difficult to explain as the rate of cleavage of A and U residues by the enzyme is similar. However, this oligonucleotide is rather susceptible to breakdown in the untreated control, giving AU and AG, which is the probable reason for the absence of the intermediate UAG.

Among the pentanucleotides studied by partial digestion with spleen phosphodiesterase, spot 41 gives only one degradation product. The M value for this degradation is 0.25 indicating a C terminal residue. The product is in the position of the marker CAG which could be taken to indicate that the sequence is CCAG. However, from the evidence of its composition and position (fig. 4.5) it cannot be CCAG but should rather be CCCCG. It will be noted from fig. 4.5 that CCCG and CAG have identical mobilities on the second dimension thus accounting for the incorrect initial interpretation. Spot 44 is another oligonucleotide which gives only one degradation product. The M value of 0.67 indicates a 5′-terminal A residue and again the product runs in the CAG position. However the reader will notice that by reference to spot 41 (which has already been discussed) one might suspect that the product was CCCG instead of CAG. The position of spot 44 on the fingerprint (not shown but in the dashed area immediately above spot 17 in fig. 4.5b) effectively distinguishes between the alternatives ACCCG and ACAG. Only the former agrees with its position on the fingerprint (see fig. 4.6). Here again confusion arose because of the slow rate of degradation of oligonucleotides containing C residues. Other examples of oligonucleotides which degrade incompletely are spots 46, 61, 63 and 89 (fig. 4.9a) all of which give as a terminal product CCG which is uniquely identified from its position. These sequences may be deduced from the M values of the degradation products which may be measured from the results shown in fig. 4.9a and checked by reference to appendix 6.

Brownlee and Sanger (1967) and Forget and Weissman (1968) have both stressed that for the analysis of the *larger* oligonucleotides from 5S RNA the sequence cannot usually be read off from the M values because of spurious splits occurring in the control or due to contaminating enzyme activity. It was therefore necessary either to elute the degradation products in order to establish their composition or to use other methods for determining the sequence of the oligonucleotide (see ch. 5).

4.6.2. *Nucleotides from pancreatic ribonuclease digests*

The degradation method with spleen phosphodiesterase gives particularly clear results when applied to oligonucleotides derived from pancreatic ribonuclease digests of rRNA. Pancreatic oligonucleotides have the general composition of $(A_nG_n)C$ or $(A_nG_n)U$ where n can be 0 or any integer – 1, 2, 3, etc. Therefore in studying products derived by spleen phosphodiesterase digestion of these oligonucleotides it is only necessary to ascertain whether an A or a G residue is split off. Moreover, as these residues appear to be cleaved at similar rates, a complete spectrum of degradation products is usually obtained. This is in contrast to the difficulty experienced with C residues in studying T_1-oligonucleotides by this method (§ 4.6.1). Thus, the sequences of pentamers and some hexamers present in rRNA were easily obtained by this method by using the M values for A and G quoted in table 4.1. It should be noted that although the range of M values for A (0.4-1.1) almost overlaps that for G (1.2-3.5) there is rarely any confusion. This is because the extreme values for A (i.e. 1.1) and for G (1.2) are only observed for degradation of the very small oligonucleotides, which in any case are best identified from their position on the DEAE-paper. Examples of these extreme values (taken from fig. 4.8) are GAC → AC (M value for G = 1.2) and AGU → GU (M value for A = 1.1). In most cases the M value for A is approximately 0.5 and for G approximately 2.

Some examples of partial spleen phosphodiesterase digestion of end products of pancreatic RNase digestion are shown in fig. 4.9b. Nucleotide 41 (from fig. 4.8b) gives bands in the positions shown on further partial digestion with spleen phosphodiesterase. The M value for the first cleavage is 0.46 indicating a 5'-terminal A residue (table 4.1). The two following splits have M values of 0.68 and 1.8 indicating that the next two residues are A and G, respectively. The product of the third split is identified by its position as it corresponds to GC in the pancreatic RNase digest of rRNA run as a marker. As spleen phosphodiesterase splits off mononucleoside 3'-phosphates from the 5'-end of an oligonucleotide the sequence AAGGC may be deduced. As this analysis depends solely on mobility values, it should be confirmed by showing that a T_1-RNase digestion gave AAG, G and C (§ 4.4.2.2). Difficulties can arise in the interpretation of the results of Fig. 4.9b if mixtures are present. Thus spot 47 gives bands in the position of both AC and GC in the marker digest indicating that at least two different sequences are present. In such a case a knowledge of the composition, which is $(A_2G_2)C$ (fig. 4.8b) is useful for interpreting the results. The first two degradation steps for spot 47 indicates the sequence GA (M values of 1.7 and 0.52, respectively). However a possible third degradation has an M value of 0.1 which would indicate a C residue. This is clearly impossible as C residues can only be 3'-terminal in complete pancreatic RNase products. The two bands running close together, however, do correspond to the positions of AGC and GAC in the marker digest. This interpretation would fit with the presence of the two dinucleotides GC and AC, already noted. The sequence of the two components of spot 47 is thus GAGAC and GAAGC. The only other point to notice is that the presence of a single band for the first degradation product for the two different sequence of the two components of spot 47 is thus GAGAC and reflects the fact that AGAC and AAGC have the same mobility. It should again be emphasised that these sequences, derived entirely by inspection of a radioautograph, must be confirmed by studying complete T_1-RNase digests of the oligonucleotide in question.

Spot 43 is more difficult to interpret. From the M values of the successive splits (1.0, 0.87, 0.45, 0.62) and the presence of the dinucleotide GC, the sequence could be AAAAGC. However, this result would not fit with its composition which is $(A_2G_2)C$ (fig. 4.8b). Without a knowledge of the intensities of the various bands shown in fig. 4.9b this is difficult to interpret correctly. However, with the information that the third band from the origin (corresponding to the position of GU in the marker digest) is weaker on the radioautograph than either the second or the fourth band, one might suggest that this represented a minor or contaminating band. Ignoring this third band, the sequence may be read off from the M values (1.0, 1.73, 0.62) as AGAGC. This third band is in a position (given that the oligonucleotide contains C and not U as its end group) of GAAC so that it is possible that the sequence GGAAC may be present as a contaminant. This hypothesis,

TABLE 4.2

Isomers of composition $(A_2G_2)C$ in P-RNase digests of rRNA.

Isomer	Spot no. (fig. 4.8)
AAGGC	41
GAGAC GAAGC	47
AGAGC	43
GGAAC AGGAC	not found

however, was proved to be incorrect by studying the products of T_1-RNase digestion of spot 43 which gave only C and AG but no AAC or G. Spot 43 is therefore AGAGC, and GGAAC is absent. The reader will notice that, of the six possible isomers with the composition $(A_2G_2)C$, only four were in fact detected (table 4.2). The two isomers which were not found (GGAAC and AGGAC) are presumably absent, or in low yield, in rRNA. The analysis of spot 66 is straightforward from the results of fig. 4.9b and is left for the reader to sequence. The result may be checked by reference to appendix 6.

It may thus be concluded that partial digestion with spleen phosphodiesterase is particularly useful for studying the sequence of oligonucleotides derived by pancreatic RNase digestion. It has been successfully used by Brownlee and Sanger (1967) and Forget and Weismann (1968) in the study of 5S RNA as well as by Min-Jou and Fiers (1969) in studying MS2 RNA sequences.

4.7. Conclusion

The fractionation procedure described in this chapter is capable of distinguishing the di, tri- and most of the tetranucleotides in digests prepared by the action of T_1-RNase or pancreatic ribonuclease on mixed 16S and 23S rRNA. The resolution of the larger oligonucleotides depends on the complexity of the mixture. It has proved a useful tool in the study of sequences in small RNA's as is amply illustrated in the next chapter as well as in purified 16S RNA (Fellner et al. 1970). Most of the information required about a mixture of oligonucleotides may be obtained by running it on the two-dimensional system using 7% formic acid for the DEAE-cellulose dimension. The approximate composition of a spot is determined from its position and the sequence may be determined by partial degradation with spleen phosphodiesterase and other procedures (see also § 5.2.2). A comparison of the patterns obtained for the T_1-RNase digests of the 16S and 23S components of ribosomal RNA shows significant differences in the tetranucleotides and very different patterns in the areas of the larger nucleotides (Sanger et al. 1965) thus illustrating that the method can be used as an effective 'fingerprinting' procedure for RNA molecules of up to 3000 residues.

CHAPTER 5

Sequence of 5S RNA

5.1. Introduction

The fractionation and sequence methods described in ch. 4 are capable of resolving the sequence of oligonucleotides isolated from high molecular weight RNA. As an illustration of the applications of these methods for the complete sequence determination of a low molecular-weight RNA molecule, this chapter outlines the problems presented by, and the techniques required for, the determination of the complete sequence of 5S RNA of *E. coli*. This RNA has a molecular weight of 36000 and forms an essential part of the ribosome, but its exact function is unknown. The choice of 5S RNA for detailed study was influenced by its ease of purification by acrylamide gel electrophoresis (appendix V.1.2) and its high yield relative to a specific aminoacyl tRNA, where purification is a major problem (see Matthews and Gould, in prep.).

The general approach for sequencing an RNA of the size of 5S RNA will be considered first in outline. The work falls clearly into two stages, the first being the isolation and sequence determination and yield of the *end products* (or *complete* products) of T_1- and pancreatic RNase digestion. These are usually isolated by means of the two-dimensional fractionation system described in § 4.3 using ionophoresis on cellulose acetate at *p*H 3.5 in the first dimension and ionophoresis on DEAE-paper using 7% formic acid in the second dimension. The second stage involves the overlapping of these T_1- and pancreatic end products which is usually achieved by isolating *partial* digestion products with either of the above enzymes. By using low ratios of enzyme to substrate and by carrying out the digestion at 0 °C these enzymes show quite a marked, but not absolute specificity, for cleavage at single-stranded, non-helical regions of the molecule. Thus partial fragments are liberated in good yield from the more stable double-stranded regions. One of the basic problems in studying such partial digests is the difficulty in fractionation of the complex mixture of products present. This problem was largely solved, for the radioactive work, by the development of 'homochromatography' either on DEAE-paper or on DEAE-cellulose thin layers. This method, when combined with ionophoresis on cellulose acetate in a two-dimensional system, provided a powerful second fractionation system for oligonucleotides in the size range of 10 to approximately 50 residues long (§ 5.5). This *second* system is complementary to the procedure already described in § 4.3 as it is most effective for the fractionation of the *larger* oligonucleotides whilst the former is better for the smaller oligonucleotides of less than 10 residues. Together the two methods are the real basis of the entire radioactive methodology for sequencing ^{32}P-RNA by degradative methods. The analysis of firstly the smaller partial fragments, followed by the isolation of progressively larger partial fragments up to the size of, if possible, half molecules, will usually allow an unambiguous derivation of the sequence.

The following is an account of the actual results of work on 5S RNA (Brownlee and Sanger 1967; Brownlee et al. 1968; Brownlee and Sanger 1969). In §§ 5.2 and 5.3 below is described the isolation of the T_1- and pancreatic RNase *end products* of 5S RNA using the standard two-dimensional ionophoretic fractionation system already described in § 4.3. The method of partial digestion with snake venom phosphodiesterase (§ 5.2.2. below) is used for sequencing the larger T_1- end products. These end products and their yields (§ 5.4) form the basis of the structural work. Nevertheless, it is not possible at this stage to deduce any overlaps from the results in the two different (T_1- and pancreatic RNase) enzymatic digests. Overlapping is achieved by the isolation of products of *partial* digestion with T_1- and pancreatic RNase using the second fractionation procedure. This uses homochromatography either on DEAE-paper or on DEAE thin layers (§ 5.5). Enzymes other than T_1- and P-RNase may also be used to obtain fragments of the molecule and partial digestion with spleen acid RNase is described (§ 5.6.4). In addition two rather different methods of obtaining partial fragments involving reaction of the 5S RNA with blocking reagents are described (§§ 5.7 and 5.8). By a combination of these various methods* a sufficient number of partial fragments were obtained to allow the derivation (§ 5.9) of the complete sequence of 5S RNA. The chapter ends (§ 5.10) with a discussion about the possible significance of the sequence thus deduced.

For the convenience of the reader studying this chapter in detail the sequence of 5S RNA is reproduced in fig. 5.1 here. It is repeated in the form of a fold-out in this manual and the reader should refer

```
1                    10                   20                    30
pU-G-C-C-U-G-G-C-G-G-C-C-G-U-A-G-C-G-C-G-G-U-G-G-U-C-C-C-A-C-
                     40                   50                    60
C-U-G-A-C-C-C-C-A-U-G-C-C-G-A-A-C-U-C-A-G-A-A-G-U-G-A-A-A-C-
                     70                   80                    90
G-C-C-G-U-A-G-C-G-C-C-G-A-U-G-G-U-A-G-U-G-U-G-G-G-G-U-C-U-C-
                    100                  110                   120
C-C-C-A-U-G-C-G-A-G-A-G-U-A-G-G-G-A-A-C-U-G-C-C-A-G-G-C-A-U
                                                              OH
```

Fig. 5.1. The nucleotide sequence of 5S RNA of *E. coli*. In strain CA265 both C and A occur in position 12, in approximately equal yields; in strain MRE600 both G and U occur in position 13, in approximately equal yields, and the residue at position 12 is always C.

---

* As the sequence of 5S RNA was the first to be established by radiochemical methods, all these various different methods were used in order to obtain *partial* fragments. In subsequent investigations on other molecules (e.g. studies on 6S RNA (Brownlee 1971)) partial digestion with enzymes has proved to be a sufficient method (when combined with isolation of the fragments by thin-layer homochromatography) to allow a complete derivation of the sequence. The chemical blocking methods (§§ 5.7 and 5.8) were not required. Nevertheless these methods are included here, as fragments were derived by using these techniques which were essential for the derivation of the full sequence of 5S RNA. Without them, a more extensive analysis, especially of the smaller partial fragments in T_1- and P-RNase digests, would have been necessary.

to this for clarification if there is any difficulty in following the argument developed in this chapter. The reader is advised in particular to have the structure open before him while reading the section on the derivation of the sequence (§ 5.9). In this section, where reference is made to a residue number, this refers to the numbered position in the sequence (starting No. 1 at the 5'-end) in the final structure. This approach has been adopted here for clarity but the reader should note that at all stages the deduction of the sequence is a logical process and is done without reference to the final structure. An important general point is that often more experiments are performed and more data are obtained than is strictly necessary for the logical deduction of the sequence. In this way any errors in observation or interpretation are detected. For the same reason observations are usually repeated.

It is convenient to use a system of abbrevations for the many fragments isolated from 5S RNA. Thus fragments are identified by letter and number: t and a are *end products* of T_1- and pancreatic RNase digestion; T, A and B are *partial products* of T_1-, pancreatic and spleen acid RNase digestion, respectively; M and C are fragments from the partial methylation and carbodiimide procedures.

5.2. Complete T_1-RNase digestion

Fig. 5.2 shows a two-dimensional fractionation of a T_1-RNase digest of 5S RNA and a map which shows the system of numbering of the spots. This fractionation includes all the T_1-nucleotides except spot 19 (the 3'-terminal nucleotide CAU_{OH}). Fig. 5.3 shows a two-dimensional fractionation of a combined T_1-RNase and alkaline phosphatase digest and a map which shows the position of dephosphorylated nucleotides and their sequences. These sequences are shown on these plates for the convenience of the reader. The methods used to derive them will now be discussed.

Preliminary information about the composition of the nucleotides can be deduced from their position on the 'fingerprint' by reference to fig. 4.7a, which shows the relationship between the composition of a nucleotide from a T_1-RNase digest and its position. A similar predic-

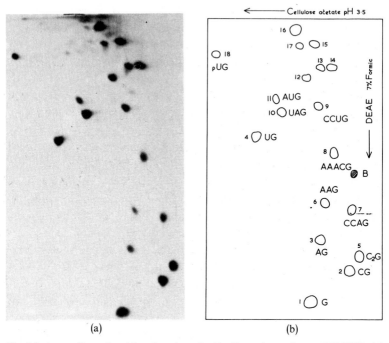

Fig. 5.2. A two-dimensional fractionation of a T_1-ribonuclease digest of 5S RNA. (a) Radioautograph and (b) a map of the major spots with their identifying number and the sequence of the smaller nucleotides. B is the blue marker. Spot 19, CAU_{OH}, was not included in this fingerprint – it is the slowest nucleotide in the first dimension but the fastest nucleotide in the second dimension.

tion for the dephosphorylated nucleotides (fig. 5.3) can be deduced by reference to fig. 4.7b. The sequence of the smaller nucleotides (tetranucleotides and smaller) was worked out by alkaline hydrolysis, by digestion with pancreatic RNase and by partial digestion with spleen phosphodiesterase as described in ch. 4. The sequence of the larger nucleotides (pentanucleotides and larger) was determined in two stages. Table 5.1 summarises the results of the first stage which includes the identification of the 5'-terminal residue by digestion with venom phosphodiesterase (§ 4.4.2.4) and the products of pancreatic RNase

5S RNA

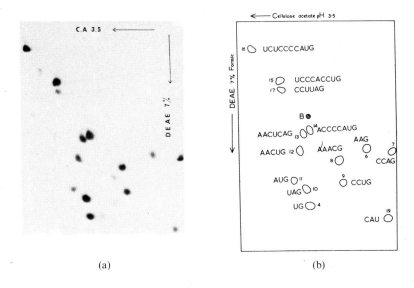

Fig. 5.3. A two-dimensional fractionation of a combined T_1-ribonuclease and bacterial alkaline phosphatase digest of 5S RNA. (a) Radioautograph and (b) a map of the major spots with their identifying numbers and sequence. CG, CCG and AG were not included in this fingerprint. Their position can be seen by referring to fig. 4.7b. B is the blue marker. It should be noted that all sequences in this digest have a 3'-terminal hydroxyl group.

digests (§ 4.4.2.1) as well as the products of 'secondary splits' or second T_1-RNase products (see § 5.2.1 below). The results of these methods* were sufficient to deduce the sequence of all the nucleotides except nos. 13, 16 and 17. An attempt at the determination of their sequence by partial digestion with spleen phosphodiesterase, which had proved useful for sequencing *smaller nucleotides* isolated from 16S and 23S

---

* Two other sequence methods, which were not available at the time of this work in 1966 but which are recommended now (1970) are described in ch. 6 in the discussion of the special problems that occur in sequencing even longer T_1-RNase end-products than were encountered in the work on 5S RNA. The T_1-end product may be digested with RNase U_2 (§ 6.6) which cleaves specifically at A residues. This method supersedes the method of 'secondary splitting' described in this chapter (§ 5.2.1). The method of P-RNase digestion after carbodiimide-blocking (§ 6.5) is also useful as it gives specific cleavage at C residues.

Subject index p. 261

TABLE 5.1

Nucleotides from ribonuclease T_1 digests.

Spot no. (figs. 5.2 and 5.3)	Composition* Alkaline digest of nucleotides from ribonuclease T_1 digests				Venom phosphodiesterase digest of nucleotides from combined ribonuclease T_1 and phosphatase digests				5'-terminal residue deduced	Pancreatic ribonuclease products**	Second ribonuclease T_1 products (table 5.2)	Structure deduced
	C	A	G	U	C	A	G	U				
8	1.1	2.9	1.0	—	×	×	×	—	A	AAAC, G	—	AAACG
12	1.1	2.1	1.0	1.0	×	×	×	×	A	AAC, U, G	—	AACUG
13	2.1	3.0	1.0	1.1	2.2	1.8	1.0	1.0	A	C(1.3), AAC(1.0), AG(1.0), U(1.2)	(C_2A_2U), AG	AAC(UC)AG
14	3.9	2.2	1.0	1.1	3.4	1.2	1.0	1.0	A	C(3.1), AC(10), G(1.1), AU(0.9)	(AC), AUG	ACCCAUG
15	4.6	1.1	1.0	2.0	4.7	1.0	1.0	1.0	U	C(3.8), AC(1.0), G(1.3), U(1.9)	(C_3U), ACCUG	ACCCCACCUG
16	4.8	1.0	1.0	3.0	5.1	0.9	1.0	1.9	U	C(4.9), G(1.0), AU(1.0), U(2.0)	$(C_{4-5}U_2)$, AUG	UCCCACCUG
17	2.0	1.0	1.0	1.9	×	×	×	×	C	C(2.5), AG(1.0), U(2.4)	(C_2U_2), AG	$U(C_5U)$AUG
18	—	—	1.0	—pUp†(0.9)	—	—	—	—		—	—	C(CU₂)AG
19	×	×	×	—††	—	×	—	×	C	—	—	pUG
												CAU_{OH}

* The compositions are expressed as the molar yield of a given nucleotide relative to one mole of G and were estimated in a scintillation counter. Where the results are given as '×'s' instead of figures, the composition was estimated by visual inspection of the radioautograph. The figures are an average of at least three separate experiments.

** The yields are expressed as moles relative to 1 mole of the dinucleotide released in each case.

† Its mobility relative to U = 1 on paper ionophoresis at pH 3.5 is 1.64.

†† Its composition, found by a complete digest with venom phosphodiesterase, was (A, U), thus showing it has a 3'-terminal U residue with a fr∴ 3'-hydroxyl group.

ribosomal RNA as described in § 4.6.1 gave ambiguous results when applied to the larger nucleotides 13, 14, 15, 16 and 17 from 5S RNA, probably because 'secondary splits' were occurring.

5.2.1. Secondary splits

This procedure caused a fairly specific cleavage of T_1-oligonucleotides at Py-A bonds and was achieved by incubating the oligonucleotide with 10 μl of 0.1 mg/ml T_1-RNase in 0.01 M Tris-chloride, 0.001 M EDTA, pH 7.5 for 30 min at 37 °C. Fractionation was carried out by ionophoresis on DEAE-paper using the pH 1.9 system until the blue marker (appendix 2) had run halfway. A T_1-RNase digest of 5S RNA was run as a marker to indicate mobilities. The products of secondary splitting were analysed for their composition or by reference to their position. Fig. 5.4 shows the 'secondary splits' of the larger nucleotides and the main products of the split, labelled a and b, are collected in table 5.2. Information from both the composition and the position of

TABLE 5.2

Products of 'secondary splits' of nucleotides from ribonuclease T_1 digests.

Band no. (fig. 5.4)	Composition*				Probable structure
	C	A	G	U	
13a[†]	1.9	1.8	—	1.0	(C_2A_2U)
b	—	×	×	—	AG
14a	—	×	×	×	AUG
b	3.8	1.0	—	—	(C_4A)
15a	2.0	1.0	0.7	0.8	ACCUG**
b	3.4	—	—	1.0	(C_3U)
16a[††]	4.3	—	—	2.0	$(C_{4-5}U_2)$
b	—	1.0	1.0	1.0	AUG
17a	2.0	—	—	2.0	(C_2U_2)
b	—	×	×	—	AG

* The yields were expressed relative to the residue shown in 1.0 or 2.0 amounts.

** Found by partial digestion with spleen phosphodiesterase.

[†] Average of three separate experiments.

[††] Average of two separate experiments.

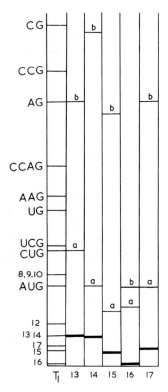

Fig. 5.4. Diagram showing the 'secondary splits' of some T_1-ribonuclease nucleotides fractionated by ionophoresis on DEAE-paper at pH 1.9. T_1 is a marker T_1-RNase digest of 5S RNA with the products identified by number (see fig. 5.2) or sequence. The thick line shows the position of unchanged nucleotides.

the nucleotide on the ionophoresis was used in determining their probable structure. For this purpose it is useful to remember that a U residue behaves much like a G residue and thus (C_3U) is in much the same position as (C_3G), and (C_2U_2) in a similar region to $(C_2U)G$. The extent of 'secondary splits' varied from experiment to experiment and also depended on the oligonucleotide in question. In all cases studied here the split occurred on the 5'-terminal side of an A residue and liberated an oligonucleotide with a 3'-terminal phosphate. Usually only a small proportion of the material remained unchanged. The basis of this type of cleavage is unclear. It appears, in fact, that it is not a property of T_1-RNase itself as a similar degradation is also observed if the enzyme is omitted. One possibility is that it is a partial alkaline hydrolysis which occurs during elution with triethylamine carbonate pH 10.0, although if this were true the specificity of cleavage is perhaps rather surprising.

5.2.2. *Partial digestion of large oligonucleotides with snake venom phosphodiesterase*

This method was devised by Holley et al. 1964, and the basis of the procedure has already been described in detail for unlabelled RNA in § 2.6.1.6. It will be recollected that this enzyme is an exonuclease liberating mononucleoside 5′-phosphates sequentially from the 3′-hydroxyl end of an oligonucleotide. In applying this method to radioactive oligonucleotides one of the aims was to devise a fractionation method that would separate oligonucleotides adequately on paper as well as allow the determination of the sequence directly by inspection of the mobility values of the products, as is possible in the analysis by partial digestion with spleen phosphodiesterase (§ 4.6.1). Initially the partial digest was analysed by ionophoresis on DEAE-paper at pH 1.9 and the composition of all the products determined. However, there were certain disadvantages in the fractionation at this pH as many of the fragments had a relatively high mobility near the mononucleotides. Moreover, it was not possible to determine any sequence directly by inspection of the mobility values of the products. An analysis by ionophoresis on DEAE-paper at pH 3.5 proved to be more useful in this respect, although it was considered *necessary always to check the composition of the partial degradation products released.*

The conditions of partial digestion are rather critical and depend on the size of the oligonucleotide as well as on the enzyme activity and pH of the buffer. It is usual, therefore, to attempt a trial experiment to find the optimum time of digestion at a given enzyme concentration necessary to obtain a good spectrum of degradation products. Alternatively the trial experiment may be omitted and digestion is carried out for various times as follows. The oligonucleotide, isolated as in fig. 5.3 and lacking a 3'-terminal phosphate group, is incubated with approximately 20 μl of 0.01 mg/ml of enzyme in 0.05 M Tris-chloride (pH 8.9), 0.01 M magnesium chloride at 37 °C. At 10, 20 and 30 min equal portions are loaded from the capillary tube, in which the incubation is carried out, side-by-side onto the DEAE-paper. A control, to which no enzyme is added, is essential and is loaded next to the sample. Fractionation is carried out by ionophoresis on DEAE-

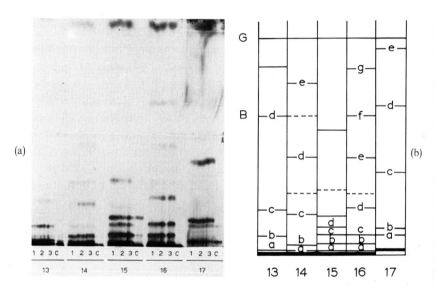

Fig. 5.5. A partial digest with snake venom phosphodiesterase of dephosphorylated nucleotides from T_1-ribonuclease and alkaline phosphatase digests, fractionated by ionophoresis on DEAE-paper at pH 3.5. (a) Radioautograph showing the degradation of nucleotides 13 to 17 after (1) 15 min, (2) 30 min, both at room temperature, and (3) 15 min at 37 °C; C is a control sample to which no enzyme was added. (b) Position of the main products, the analyses of which are given in table 5.3. A few products were not analysed, either because they were present in the control sample (----) or because they were in low yield (―――, no letter). Other products, the analyses of which are given, are not present in this experiment, but are included from other experiments where longer digestion times were used. B is the blue marker and the thick line is unchanged material. G is guanosine 5'-phosphate.

paper at pH 3.5 running the blue marker (appendix 2) approximately one third of the length of the DEAE-paper. After autoradiography the fragments produced are analysed for their composition by alkaline hydrolysis, or occasionally by further complete digestion with venom phosphodiesterase.

Fig. 5.5a shows a partial venom digest of spots 13 to 17 of fig. 5.3 fractionated by ionophoresis on DEAE-paper at pH 3.5 and fig. 5.5b a line diagram of the products obtained. The analysis of the degradation products is given in table 5.3. In the interpretation of the composition of the various products of digestion by alkaline hydrolysis it is important to remember that the 3′-terminal residue is released as a *nucleoside* and is therefore not detected. As an illustration, consider the degradation products of spot 16 in table 5.3 which are labelled a–g. Assuming that the nucleotides fractionate in order of size, where the smallest ones move the fastest, then 16g (C, U) is likely to be the trinucleotide (U, C)N where N is unknown. 16f (A, U) is obviously unrelated to the degradation product already considered and is likely to be AUG, the same product as is liberated by secondary splitting (see table 5.2). 16e (C, U) differs from 16g in having an extra U residue and is therefore (U,C)UN. Thus, 16g must be (U,C)U. Similarly 16d which analyses as (C, U), i.e. C equal to U is probably in fact (C_2U_2) and differs from 16e by a C residue. It is therefore (U,C)UCN. 16a, b and c appear to differ from 16d by having more Cs. The sequence of 16a is therefore (U,C)UCCCCN. With the additional information already deduced and presented in table 5.1 the structure of 16 can be written as UCUCCCCAUG.

Our previous experience in the analysis of partial digests of nucleotides with *spleen phosphodiesterase* on DEAE-pH 1.9 ionophoresis showed that it was possible to deduce the structure of a nucleotide directly from the mobility of its degradation products. A value of M was defined (§ 4.6.1) and its value for any one split was characteristic of the mononucleotide released. In a similar approach to this, the values of M for the 'splits' shown in fig. 5.5 on DEAE-pH 3.5 ionophoresis have been collected in table 5.4 with their values listed opposite the products of the split. The value of M is defined (as before) as equal to

TABLE 5.3

Nucleotides from partial digests with snake venom phosphodiesterase fractionated by ionophoresis on DEAE-paper at pH 3.5.

Band no. (fig. 5.5)	Composition*		Value of M (see text)	Probable structure
	Alkaline hydrolysis	Venom phospho-diesterase		
13a	C A U		2.9	A(AC)UCA
b	C A U	C A U	0.9	A(AC)UC
c	C A		1.7	A(AC)U
d	—		—	A(AC)
14a	C A		1.9	(ACC)CCAU
b	C A		2.7	(ACC)CCA
c	C A		0.9	(ACC)CC
d	C A		0.6	(ACC)C
e	—		—	(ACC)
15a	C A U		1.7	(UCCCA)CCU
b	C A U		0.9	(UCCCA)CC
c	C A U		—	(UCCCA)C
d[†]	C U		—	(UCCC)p**
16a	C U		2.5	(UC)UCCCCA
b	C U		0.9	(UC)UCCCC
c	C U		0.9	(UC)UCCC
d	C U		0.9	(UC)UCC
e	C U		0.8	(UC)UC
f[†]	A U		—	AUG
g	C U		—	(UC)U
17a[†]	C U		—	CCUUp**
b	C U		2.1	CCUUA
c	C U		1.4	CCUU
d[†]	A		—	AG
e	C & G	C U & G	—	CCU

* Compositions were estimated by visual inspection of the radioautograph and are expressed relative to the nucleotide present in lowest amounts. If a nucleotide is once underlined there are two residues, if twice underlined either three or more. It was not possible to estimate with any degree of accuracy ratios greater than 3 to 1.

** These nucleotides have a 3'-terminal phosphate; others are dephosphorylated.

[†] These nucleotides were present in the control sample, to which no venom phosphodiesterase had been added, and were the same as the products of secondary splits recorded in table 5.4.

x/y where y is the distance of any oligonucleotide from the origin of the DEAE-paper, and x is the distance between it and its first degradation product.

For the nucleotides studied it has been found that the values of M lie within the limits shown in table 5.4, *with the restriction that values of M for the 'splits' where the product of the split is faster than the blue marker are not included.* There is no overlapping between the values for C, A and U although the values for A and U are rather close to one another. Thus in no. 13 (see table 5.3) four bands, a to d were recognised of which an analysis was obtained on a, b and c. 13c (C, A) was related to 13b (C, A, U) by the loss of a U; 13b was related to 13a (C, A, U) by

TABLE 5.4

Values of M on DEAE-paper ionophoresis (pH 3.5) for nucleotides from combined ribonuclease T_1 and alkaline phosphatase digests.*

3'-terminal residue	Range of M-values	No. of observations
C	0.6–1.2	13
A	2.1–2.9	4
U	1.7–1.9	3
G	2.6–4.4	6

* The values of M for splits in which the product of the degradation is faster than the blue marker are not included.

the loss of a C, and 13a was related to unchanged material (C_2A_3U) by the loss of an A. Thus the 5'-nucleotide must be released in the following order – A, C, U – and the sequence of 13 must be (A,A,C)UCAG. This same information can be read off from the values of M for the first three splits referring to table 5.4, all of which occur proximal to the blue marker. The first with an M value of 2.9 could be either A or G, the second must be a C ($M = 0.9$), and the third a U ($M = 1.7$).

For spot 14, the values of M (see tables 5.3 and 5.4) suggested nucleotides were released in the order – U, A, C – and with the additional information from the composition of the products, the structure (A,C,C)CCAUG was deduced.

Subject index p. 261

TABLE 5.5

Summary of the evidence for the base sequence of the larger nucleotides from ribonuclease T_1 digests.

Spot no. (figs. 5.2 and 5.3)	Sequence*
13	A A C U C A G
14	A C C C C A U G
15	U C C C A C C U G
16	U C U C C C A U G

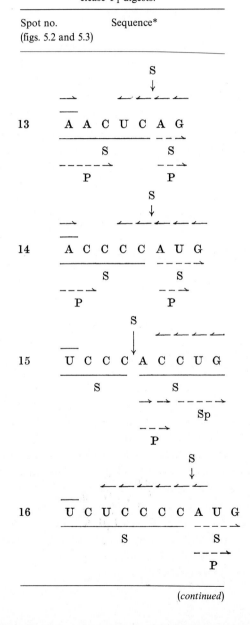

(*continued*)

TABLE 5.5 (*continued*)

Spot no. (figs. 5.2 and 5.3)	Sequence*
17	 S ↓ ←— ←— ←— ←— —— C C U U A G ————————— – – → S S

* The symbols and the method of digestion from which the sequence information was derived are as follows: (a) →, partial digestion with spleen phosphodiesterase; (b) ←, partial digestion with snake venom phosphodiesterase; (c) $\overset{S}{\underset{\downarrow}{|}}$, a secondary split at the point indicated; (d) $\overline{\underset{S}{}}$, a fragment of known composition from a secondary split; (e) $-\overline{\underset{S}{}}\rightarrow$, a fragment from a secondary split the sequence of which is known from its mobility on ionophoresis; (f) $-\overline{\underset{P}{}}\rightarrow$, a pancreatic ribonuclease fragment of known sequence; (g) ——, the 5′-terminal nucleotide as found from the difference in the composition by alkaline hydrolysis and hydrolysis with snake venom phosphodiesterase; (h) $-\overline{\underset{S_P}{}}\rightarrow$, a fragment from a partial digestion with spleen phosphodiesterase the sequence of which is known from its position on ionophoresis.

For spot 15, the values of *M* suggested nucleotides were released in the order – U,C – and knowing the composition of the products the structure (UCCCA)CCUG was deduced.

Spot 17 was particularly labile and over half was degraded in the untreated control sample to give the same products as were described under secondary splits above. Bands 17b, c and e were not present in the control. 17b (C,U) which by its position is likely to be (C_2U_2) was related to unchanged material (C_2U_2A) by the loss of an A and to 17c (C,U) by the addition of a U. 17e (C), which runs with free G, is related to 17c by the loss of another U. The 5′-mononucleotides are therefore released in the order – A,U,U – and the sequence is therefore CCUUAG.

A complete venom phosphodiesterase digest of 17e gave (C,U) and confirmed its sequence CCU.

One advantage of this method of sequence analysis compared with analysis by partial digests with spleen phosphodiesterase is that all possible degradation products occur. This must mean that the different 5′-mononucleotides are released at fairly similar rates by venom phosphodiesterase, whereas with spleen phosphodiesterase C residues are released much more slowly than the others (Sanger et al. 1965). Table 5.5 is an attempt to summarise the evidence for the nucleotide sequence of the larger nucleotides from T_1-RNase digests. This type of notation is derived from that used in sequence analysis of peptides and is a useful way of emphasising the variety of methods which can be used. Nevertheless, it is impossible to summarise completely all the evidence for a particular structure which can be checked by the reader only by careful inspection of that part of the tables which relates to the oligonucleotide in question.

The fractionation procedure for analysing the products of partial venom phosphodiesterase digestion may be modified where necessary. Thus fractionation using DEAE-pH 1.9 or DEAE-7% formic acid may be necessary for the separation of larger fragments than are encountered in 5S RNA which remain at the origin of the pH 3.5 system but which have a significant mobility on the more acid DEAE-systems (Goodman et al. 1970). Min-Jou and Fiers (1969) have suggested another modification to the method so that it is suitable for oligonucleotides with a 3′-terminal phosphate group thus avoiding the need for the combined T_1-RNase and phosphatase digestion of the total RNA. They heated the nucleotide to 90 °C for 3 min in 0.01 M Tris-chloride, pH 9.0 and then transferred to 37 °C and after 2 min added alkaline phosphatase using 1 part of enzyme to 10 parts of substrate. After 30 min at 37 °C, EDTA was added to 0.001 M, and the solution heated again to 90 °C for 3 min to destroy phosphatase activity. Finally, $MgCl_2$ was added to 0.002 M, and venom phosphodiesterase to obtain an enzyme-to-substrate ratio of about 1 to 10. 10-μl samples were taken up at different times (from 15-60 min) and fractionated by ionophoresis on DEAE-paper at pH 3.5 as above. This method may again be simplified by omitting the procedure to destroy phosphatase activity. The partial products are in any case resistant to phosphatase as they lack any terminal phosphate group. It may, in fact, be an advantage to have phosphatase present to degrade the released mononucleoside 5'-phosphates, which often have a rather similar mobility to any dinucleoside monophosphates present in the digestion mixture.

5.2.3. End groups

Spots 18 and 19 (figs. 5.2 and 5.3) occupied unusual positions on the fingerprint in that they did not lie within the usual grouping of spots, referred to as the graticule (§ 4.5.1).

Spot 19 gave Cp and Ap but no Gp on analysis with alkali (see table 5.1) and was susceptible to complete digestion with venom phosphodiesterase when it gave pA and pU. These results are interpreted as follows. The absence of a G in the alkaline digest indicates that the oligonucleotide is derived from the 3'-terminus of the molecule and thus has the form $(C,A)N_{OH}$ where N_{OH} is any nucleoside. The relative order of the C and A are deduced from the products of venom phosphodiesterase digestion which gives nucleoside 5'-phosphates (pA,pU). The new residue (pU) in this digest compared with the alkaline digest must be 3'-terminal giving the sequence $C-A-U_{OH}$. The 5'-terminal residue is therefore C which is confirmed by its absence in the venom digestion. In summary the 3'-terminal oligonucleotide of any RNA molecule is easily recognised by the lack of a G in the alkaline digest. The absence of a 3'-terminal phosphate usually makes such 3'-end groups run rather *slowly* towards the anode, or even towards the cathode in the first dimension.* The position in the second dimension

---

* Care must be taken, therefore, to include *all* the first dimension, including any material running 'backwards' in any 'fingerprint' designed to observe the 3'-end group of a molecule. For example, the oligonucleotide CCA_{OH} encountered in T_1-RNase digests of some tRNAs is *basic* at pH 3.5 with a net charge of about half positive. It therefore runs 'backwards' towards the cathode (negative electrode) on cellulose acetate at pH 3.5. At pH 1.7 (in 7% formic acid) it has about one positive charge. On DEAE-paper ionophoresis in 7% formic acid it is therefore considerably retarded, due to the ionophoresis, relative to the fastest moving nucleotide, and runs about half the rate of Gp.

is unpredictable as it depends on the terminating nucleo*side*. If this is uridine one can say, to a close approximation, that the position occupied is that of the same nucleotide without the terminal *uridine*, i.e. C-A- is the same as C-A-U_{OH}. However, if C or A are terminal nucleosides then even oligonucleotides with two Up residues will move faster than the blue marker in the second dimension. Thus the nucleotide A-U-U-C_{OH}, from 6S RNA of *E. coli* (Brownlee 1971), moves in approximately the same position on the fingerprint as C-A-G-.

Spot 18 on alkaline hydrolysis gave Gp and a product which moved faster than any known mononucleotide on paper ionophoresis at pH 3.5 (see table 5.1) thus suggesting a diphosphate. This may be identified as pUp as described in § 8.2.1 by re-running on DEAE-paper at pH 1.9. The structure of spot 18 is therefore pU-G-. Another method of clearly showing the 5'-terminal nucleotide to be Up, is to degrade the presumed pU-G- with alkaline phosphatase removing both the 5'- and the 3'-phosphate esters and giving U-G. On treatment with alkali this gives Up. In general 5'-terminal sequences, in contrast to 3'-terminal sequences, move rather fast on the standard two-dimensional fractionation system because of the extra negative charge of the 5'-phosphate group. Thus pGp, a common end group in tRNA, runs ahead of U-G- in the first dimension but has the same mobility in the second dimension. Another point to notice is that the terminal phosphate is *equivalent*, for the purposes of predicting mobility in the second dimension to adding another *U residue*, i.e. pU-G- lies in the U_2G graticule (fig. 4.7a).

5.3. *Complete pancreatic RNase digestion*

Fig. 5.6a shows the fractionation of pancreatic RNase digest of 5S RNA on the two-dimensional system using pH 1.9 in the second dimension. Like the T_1-RNase digests there is no difficulty in separating all the products in a pure form even in a fairly short separation. The results of the sequence analysis are given in fig. 5.6b. A discussion of how to sequence the smaller products in a P-RNase digest has already been given in § 4.5.2. In fact, all of the simpler nucleotides had pre-

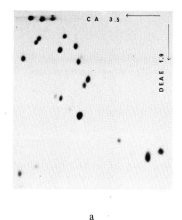 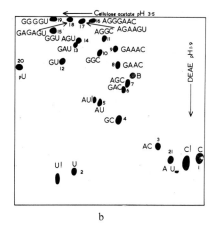

Fig. 5.6. A two-dimensional fractionation of a pancreatic ribonuclease digest of 5S RNA. (a) Radioautograph showing all the larger products except GGU, GGGGU and pU, which have run too fast to be included in this fingerprint. (b) Positions of all the major nucleotides with their identification numbers and sequences. Some cyclic nucleotides are present and are indicated by !; B is the blue marker, and spot No. 21 has a free 3′-hydroxyl group.

viously been found and analysed by alkaline hydrolysis and by digestion with T_1-RNase in digests of high molecular weight ribosomal RNA. This discussion will be restricted to the larger products not previously encountered. Table 5.6 summarises the results of analyses by alkaline hydrolysis and by T_1-RNase digestion. *Spots 20 and 21* were distinguished from all the other oligonucleotides in that they did not give either Cp or Up among the alkaline degradation products. They therefore derive from the two ends of the molecule already deduced from the T_1-RNase digestion (§ 5.2.3) and need not be discussed further. *Spot 19* gave only G's and a U. Although several attempts were made at determining the quantitative ratio of the two nucleotides it was not possible to decide whether the ratio was nearer 3:1 or 4:1 as figures between these values were obtained. However, examination of the *position* of spot 19 on fig. 5.6 clearly distinguished these possibilities. Thus if it were G_3U it would have to occupy a position

Subject index p. 261

TABLE 5.6
Nucleotides from pancreatic ribonuclease digest.

Spot no. (fig. 5.6)	Composition*	Products from ribonuclease T_1**	Structure deduced
16	A G C	AAC, AG, G	(AG,G,G)AAC
17	A G U	AG, AAG, U	(AG,AAG)U
18	A G U	AG, G, U	(AG,AG,G)U
19	G U		$G_{3-4}U$
20	pU		pU
21	A		AU_{OH}

* See footnotes to table 5.3.

** The yield of di-or trinucleotides is expressed as a molar yield. Thus in 16, although the intensity of blackening of the radioautograph by G and AG is similar, the molar yield of G is twice AG.

about one third the rate of G_2U (spot 15). This follows from the M value for G which is approximately 2.0 for ionophoresis on DEAE-paper at pH 1.9 (table 4.1). Clearly, spot 19 is much slower than this and must be *at least* G_4U from its position. This example illustrates that the position of oligonucleotide on the fingerprint may be a more reliable indication of composition than an actual analysis. In any case, it is always a valuable check. A possible reason for the *low* analysis for G is that purine mononucleotides are more susceptible to slight further degradation during alkaline hydrolysis to nucleotides and phosphate ($R_u = 1.7$ on paper ionophoresis at pH 3.5) than are pyrimidine nucleotides. Inorganic phosphate is, in fact, usually observed in low yields in alkaline hydrolysates. Another unambiguous method of determining the number of G residues is to observe the degradation pattern with partial digestion with spleen phosphodiesterase as described below (§ 5.3.1). *Spot 16-18* could not be sequenced from the results shown in table 5.6 and the further procedure of partial digestion with spleen phosphodiesterase (§ 5.3.1) below was required.

5.3.1. *Partial digestion with spleen phosphodiesterase*

This method has been described in detail in § 4.6.2 and fig. 5.7 shows the

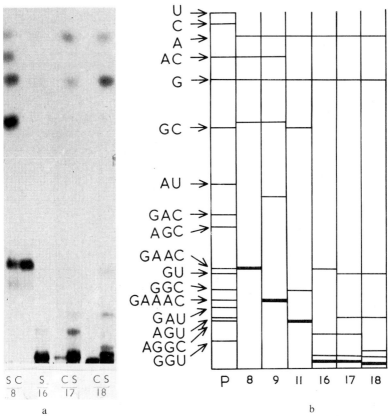

Fig. 5.7. A partial digest with spleen phosphodiesterase of certain pancreatic nucleotides fractionated by ionophoresis on DEAE-paper at pH 1.9. (a) Radioautograph: S is the sample and C the control. (b) Position of the partial products, with the unchanged material shown by a thick line. P is a pancreatic ribonuclease digest of 5S RNA run as a marker.

degradation of spots 16-18 using the method (which removes nucleoside 3'-phosphates sequentially from the 5'-end). By inspection of fig. 5.7, M values for the various degradations may be calculated indicating (see table 4.1) the mononucleotide released at each step. Thus, for no. 16 the values of M for the various steps are 0.8 (A), 2.3 (G) and 2.0 (G), with the last step giving rise to a band in the position of GAAC in the P-RNase digest of 5S RNA run as a marker. No. 16 is, therefore, AGGGAAC.* The sequence of nos. 17 and 18 is quite straightforward and may be deduced directly from the M values as AGAAGU and GAGAGU, respectively. The degradation of no. 19 is not shown but the values of M for the first two steps are 2.7 and 3.0, the latter step giving a band in the GGU position. No. 19 is, therefore, GGGGU. These results were in agreement with the results of table 5.6. In summary, the ease of analysis by the degradation with spleen phosphodiesterase made the complete sequence analysis of the pancreatic ribonuclease products much quicker than the analysis of nucleotides from T_1-RNase digests.

5.4. Relative molar yields and discussion

Table 5.7 shows some estimates of the relative molar yields of the nucleotides from both the T_1-RNase and pancreatic ribonuclease digests in figs. 5.2 and 5.7. These were calculated as follows. After fractionation on the two-dimensional system the radioactivity of each nucleotide was estimated in a liquid scintillation system by direct measurement of the counts per minute in that area of paper which contained the nucleotide. A circle of paper containing the oligonucleotide was cut out and was partially rolled up into a glass vial (2 inch long $\times \frac{1}{2}$ inch diameter) which fitted in a glass scintillation vial. The vial was filled with scintillant (0.4 % BBOT in toluene), and counted in a liquid scintillation counter (appendix 3) at room temperature. A circle of paper from an area of the ionophoretogram where no radioactivity was present, served as a control for the counting proce-

* Of course, larger products than heptanucleotides may be encountered in P-RNase digests of molecules other than 5S RNA and some modification of the method of partial digestion with spleen phosphodiesterase is necessary to sequence them. In particular, fractionation may be carried out using ionophoresis on DEAE-paper in 7% *formic acid* to improve the mobility of the larger fragments. An alternative approach is to use partial digestion with venom phosphodiesterase (§ 5.2.2) again using DEAE-ionophoresis in 7% formic acid for the fractionation of the larger products. Both of these methods have been used by Min-Jou and Fiers (1969) in their studies on large oligonucleotides, the largest being GAGGAGAAGC from the MS2 bacteriophage RNA.

dure. Normally at least 2000 counts were collected. No correction for quenching was made. The counts per minute divided by the number of phosphate residues in the nucleotide in question was expressed relative to $AG = 2.0$ in T_1-RNase digests and to $AU = 2.0$ in pancreatic RNase digests. These dinucleotides were taken as *internal standards* because there was evidence from the sequence information present in the other digest that this should represent the actual molar yield. Thus, AG in a T_1-RNase digest derives from a GAG sequence in the nucleic acid. The only two GAG sequences appear in the same pancreatic nucleotide, GAGAGU. Similarly the 2 AU residues in a pancreatic RNase digest are present in nucleotides 14 and 16 of the T_1-RNase digest.

To test the reproducibility of a two-dimensional fractionation system, aliquots of a T_1-RNase digest were fractionated in three different ways. In experiments 1 and 2 the digest was fractionated, on one dimension only, by ionophoresis on DEAE-paper at pH 1.9; in experiment 3 fractionation was on the usual two-dimensional system using pH 1.9 for the second dimension; and in experiment 4 the digest was fractionated on a two-dimensional system which used ionophoresis on DEAE-paper in 7.5% triethylamine carbonate (pH 9 to 10) in the second dimension. The individual yields using these various methods (table 5.7) are all within 15% of the mean value and most are much closer than this to the mean. Thus the losses, at least for the smaller nucleotides nos. 1 to 7, on transfer of material from the cellulose acetate to DEAE-paper must be negligible. Also nucleotides cannot be appreciably degraded by fractionation on DEAE-paper at pH 1.9 as the yields are similar to those obtained by fractionation on DEAE-paper at pH 9 to 10. However, the yields of the larger nucleotides (penta- to decanucleotides) were sometimes lower than 1.0. This probably reflected several factors, among them a loss on transfer from cellulose acetate to DEAE-paper as well as a low yield because the digestion with T_1-RNase had not gone to completion so that some nucleotides with cyclic phosphate were present which could not always be included in the figures given in table 5.7. Experiments 5 (the same fractionation as is shown in fig. 5.2) and 6 were separate digests of another preparation of 5S RNA fractionated using 7% formic acid in

Subject index p. 261

TABLE 5.7
Molar yields of nucleotides.*

Spot no. (fig. 5.2)	Sequence	Ribonuclease T_1 digests						Average molar yield	Nearest integer
		1**	2	3	4	5	6		
t1	G	12.8	12.8	10.4	11.2	13.7	12.6	12.2	12
t2	CG	4.3	4.2	4.3	4.4	6.1	7.8	5.2	5
t3	AG	2.00	2.00	2.00	2.00	2.0	2.00	2.00	2
t4	UG	3.5	4.0	4.0	3.4	4.7	5.0	4.1	4
t5	CCG	2.7	2.7	2.9	2.9	3.8	3.8	3.1	3
t6	AAG	1.3	1.1	1.1	1.1	0.9	0.9	1.1	1
t7	CCAG	0.7	0.7	0.7	0.9	1.3	1.4	1.0	1
t8	AAACG			0.7	0.7	0.8	0.8	0.8	1
t9	CCUG			0.7	0.8	1.1	1.3	1.0	1
t10	UAG			2.9	2.8	3.4	3.7	3.3	3
t11	AUG			0.9	1.0	1.0	1.0	1.0	1
t12	AACUG			0.8	0.7	1.1	1.0	0.9	1
t13	AACUCAG			0.6	0.8	0.7	0.9	0.8	1
t14	ACCCCAUG			0.5	0.7	0.9	0.9	0.8	1
t15	UCCCACCUG			0.6	0.7	1.1	1.0	0.9	1
t16	UCUCCCCAUG			0.4	0.6	1.2	1.3	0.9	1
t17	CCUUAG			0.3		0.5	0.5	0.4	0.5
t18	pUG			0.9	0.5	1.2		0.9	1
t19	CAU_{OH}			0.7				0.7	1
Others	U(CU)G			0.1				0.1	0
Others	CUG			0.3	0.4	0.0	0.1	0.2	0
Others	UCG			0.2	0.3	0.0	0.0	0.1	0

Total = 118

TABLE 5.7 (continued)

Spot no. (fig. 5.6)	Sequence	Pancreatic ribonuclease digests				Average molar yield	Nearest integer
		Molar yield*					
		7	8	9	10		
a1	C	17.6	20.8	18.9		19.1	19
a2	U	4.0	4.8	4.8		4.5	5
a3	AC	1.1	1.1	1.2		1.1	1
a4	GC	7.0	7.0	7.1		7.0	7
a5	AU	2.00	2.00	2.00	2.00	2.00	2
a6	GAC	1.1	1.0			1.1	1
a7	AGC	1.9	2.0			2.0	2
a8	GAAC	1.0	0.9			1.0	1
a9	GAAAC	1.0	0.8			0.9	1
a10	GGCC	2.0	1.7			1.9	2
a11	AGGC	0.8	0.6			0.7	1
a12	GU	2.1	2.8	2.4	2.3	2.4	2 to 3
a13	GAU	1.0	1.0	1.0		1.0	1
a14	AGU	1.2	1.2	1.0		1.1	1
a15	GGU			2.4	2.6	2.5	2 to 3
a16	AGGGAAC	0.4				0.4	0.5
a17	AGAAGU	1.0	0.8			0.9	1
a18	GAGAGU	1.0	0.7			0.9	1
a19	GGGGU						1
a20	pU						1
a21	AU_{OH}	0.8	0.9				1
Others	A_2C	0.05					0
							Total = 115

* Expressed relative to AG or AU taken as 2.00.

** Experiments 1 to 4 were separate fractionations of the same digest. 1 and 2 were fractionated on DEAE-cellulose at pH 1.9, 3 was fractionated on the usual two-dimensional system using the pH 1.9 mixture for the second dimension, and 4 was fractionated on a two-dimensional system which used ionophoresis on DEAE-paper using 7.5% triethylamine bicarbonate pH 9 to 10 in the second dimension. Experiments 5 and 6 were separate digests and fractionations of another preparation. Experiments 7, 8, 9 and 10 were separate digests and fractionations of a third preparation. Gaps in the table indicate that no quantitative estimate was made in that particular experiment.

the second dimension of the two-dimensional system. There were no cyclic nucleotides present whereas in other experiments (e.g. experiment 3) up to half the free G content was present as G cyclic phosphate. The yields relative to AG were about 20% higher than in experiments 1 to 4 for most nucleotides; e.g. CG, CCG, UG, CCAG, CCUG, although they were the same as in experiments 1 to 4 for AAG, AAACG and AUG. Probably the yields are high relative to AG because AG itself as well as AAG, AAACG and AUG are relatively unstable to over-digestion with T_1-RNase (or to a contaminant of it) and degrade further. This interpretation accounts for the larger nucleotides being recovered in an apparent yield which is greater than 1. If the yields are calculated relative to $UG = 4.0$ in experiments 5 or 6 then the yields of the larger nucleotides do more nearly approximate to the expected yield. Thus, the yields of 15, 16 and 17 corrected in this way are 0.9, 1.0 and 0.4, respectively.

In summary, the differences between experiments 5 and 6 and 1 to 4 are caused by the different conditions of digestion rather than by variations in yields caused by experimental error in the method of fractionation. In general it would appear that the main limitation in the use of the method for quantitative work, will be in the control of the hydrolytic procedure. This could arise from lack of specificity of the enzyme.

It is instructive to compare at this stage, these yields with the values predicted from the final structure. In making this comparison, the average of experiments 1 to 6 was used as the two opposing factors of under-digestion and over-digestion might be expected to cancel one another out. This may be seen to be true and there is, in fact, a reasonable agreement as may be seen from Table 5.8. The deviation from predicted figures is probably highest in three oligonucleotides, viz. AGGGAAC, GGU, and GGGGU in the P-RNase digest. In an attempt to improve the specificity of the digestion, which might explain some of the ambiguous yields, we digested with P-RNase at higher ionic strengths of 0.3 M as suggested by Min-Jou and Fiers (1969). We carried out the digestion in 0.01 M Tris-chloride, 0.001 M EDTA, 0.3 M ammonium acetate pH 7.5. Under these

TABLE 5.6
End products of ribonuclease digestion of 5S-RNA (MRE 600).

Ribonuclease T$_1$			Ribonuclease A		
	Yield			Yield	
Sequence	Found	Calculated from final structure	Sequence	Found	Calculated from final structure
G	12.2	11	C	19.1	19
CG	5.2	5	U	4.5	6
AG	2.0	2	AC	1.1	1
UG	4.1	4	GC	7.0	7
CCG	3.1	3.5	AU	2.0	2
AAG	1.1	1	GAC	1.1	1
CCAG	1.0	1	AGC	2.0	2
AAACG	0.8	1	GAAC	1.0	1
CCUG	1.0	1	GAAAC	0.9	1
UAG	3.3	3.5	GGC	1.9	2
AUG	1.0	1	AGGC	0.7	1
AACUG	0.9	1	GU	2.4	2.5
AACUCAG	0.8	1	GAU	1.0	1
ACCCCAUG	0.8	1	AGU	1.1	1
UCCCACCUG	0.9	1	GGU	2.5*	3
UCUCCCAUG	0.9	1	AGGGAAC	0.4**	1
CCUUAG	0.4	0.5	AGAAGU	1.1	1
pUG	0.9	1	GAGAGU	1.0	1
CAU$_{OH}$	0.7	1	GGGGU	†	1
			pU	0.9	1
			AU$_{OH}$	0.9	1

*, **, †. These figures may be corrected to 2.9, 0.6 and 0.7 respectively, see text. The absence of a figure in the 'Found' column indicates that no estimation was made.

conditions the enzyme was more active than at the lower ionic strengths previously employed, and few cyclic nucleotides were present on the fingerprint. Also, there was a slight improvement of the yield of GGU to 2.9, AGGGAAC to 0.6 and GGGGU to 0.7 (compare with table 5.7). It is clear, however, in comparing even these experimental results with the values predicted from the final structure (see table 5.8), that it is difficult to get theoretical yields of P-RNase products which contain several adjacent guanine residues. In fact, G could often be detected as a faint spot on the 'fingerprint' of a P-RNase digest. Min-Jou and Fiers (1969) also recommend re-purification of the commercially available enzyme according to the treatment of Dutting et al. (1966) as described in § 2.2.3.

Another difficulty which did not seem to be related to any lack of specificity of the enzyme was the consistently low yields from T_1-nucleotide no. 17, which was present in approximately 0.5 molar yields. In another strain of *E. coli* (strain CA 265), and in many other laboratory strains, this particular nucleotide is absent but, instead, a new oligonucleotide CAG is present in the 'fingerprint', again in approximately 0.5 molar yields. Although these low yields might have indicated the presence of a contaminating species of RNA, the absence of any other nucleotides in equivalent yields rather suggested they were part of the structure. This was clarified when these sequences were found to represent *alternative* sequences of a particular part of the molecule from the results of partial digestion.

A comparison of the sequences of table 5.8 in the two types of digest allow one to cross-check the sequences. Thus, any A_nC, A_nG or A_nU sequences must necessarily be present in equal amounts in both digests. Inspection of the sequences and their yields show, in fact, that AAC = 2, AAAC = 1, and AU = 3, in both digests. For AG there are 7.5 in the T_1-digest and 8 in the pancreatic digest. This discrepancy indicates that one of the quantitations, one might suspect the three UAG sequences, is incorrect and should be increased to 3.5. This figure, would, in fact, be nearer the average experimental figure of 3.3, although the accuracy of the quantitation is certainly not sufficient to distinguish between 3 and 3.5 as may be seen by examination of the individual variation of the figures for UAG. Other detailed comparisons of the sequence can be made. Thus the yield of U in the P-RNase digest of 4.5 (experimental value) may be seen to be too low by examining the number of pyrimidine-U sequences in the T_1-digest. This turns out to be 5.5. Similarly, the number of Py-C's may also be correlated, and in both cases was 19. The total number of residues may also be calculated at this stage and agree to within 3%. In summary, it may be said that there should be good, although not necessarily exact correlation of the results in the two digests at this stage of the work. For study of the sequences of other RNA's we would recommend that quantitations should be carried out on pure material, only if the 'fingerprints' are good and show no streaking and that the average of two separate digests and fractionations may be taken. Any discrepancies in the results are usually resolved at a later stage in the work, when the partial digests are analysed.

Although no estimate of the purity of the 5S RNA was made, the absence of any contaminant in yields greater than 0.1 molar yields suggested that the material was over 90% pure. In fact, if a random mixture of contaminants is present, it is still possible to do sequence work on much less pure material and probably a 50% purity would be sufficient. However, some experience and discretion is necessary in working with impure material and it is recommended that some material should be prepared of over 90% purity and then, as the specific end products are recognised, less pure material may be used to obtain the correct fragments in order to sequence them.

We might summarise the results of this stage in the sequence determination as follows. When the two-dimensional ionophoretic method of fractionation is applied to the examination of radioactive nucleotides from ^{32}P-labelled 5S RNA of *E. coli*, all of the 19 nucleotides in a T_1-RNase digest and all the 21 nucleotides in a pancreatic RNase digest are separated from one another; the yields of nucleotides are reproducible and approximate to the expected integral figures.

Various methods may be used in the sequential analysis of the oligonucleotides but the most generally useful is to partially degrade them with exonucleases and isolate and analyse all the products.

Thus, oligonucleotides from pancreatic RNase digests are conveniently and rapidly sequenced using spleen phosphodiesterase, and nucleotides from combined T_1-RNase and alkaline phosphatase digests are analysed using snake venom phosphodiesterase. Secondary splitting is a useful method of analysing nucleotides from T_1-RNase digests.

Although the results of the two different ribonuclease digests confirmed one another, it *was not possible to use information about the pancreatic end products to join together any of the T_1-end products by overlapping.* This is in contrast to work on proteins where the determination of the complete sequence of products from two enzyme digests will usually give the complete structure of the protein. *Nevertheless these end products form the basis of the structural studies and their sequence should be determined before proceeding to the next stage of the work described below.* At this stage in the work the 5'-terminus and 3'-terminus were, however, known; also an estimate of the total numbers of residues could be made.

5.5. 'Homochromatography' and fractionation of partial enzymatic digests

Initially Brownlee and Sanger (1968) attempted to fractionate the products of *partial* digestion with T_1-RNase on the standard two-dimensional system (§ 4.3). However, most of the large nucleotides remained at the origin of the DEAE-paper although they separated well in the cellulose acetate ionophoresis dimension. It was, thus, necessary to find an alternative fractionation method for the second dimension which would be suitable for nucleotides of this size. The best results were obtained using chromatography on DEAE-paper, rather than ionophoresis. In order to prevent the large nucleotides being absorbed too strongly on the DEAE-paper, a high concentration of an anion was necessary. This was not possible with ionophoretic systems as it increased the conductivity too much with consequent overheating. In the normal column chromatographic systems with DEAE-cellulose, the necessary anion concentration is achieved by using a sodium-chloride gradient to displace the nucleotides from the

DEAE groups. An alternative procedure, however, is to displace the radioactive nucleotides by a series of anions of different valency or affinity for the DEAE groups. Such a system is produced when a concentrated mixture of oligonucleotides (or of any other suitable anions of different valency) is applied to the end of an DEAE chromatogram. The oligonucleotides saturate the DEAE groups and displace one another producing a series of fronts. The smaller ones with lower valency or affinity are displaced by the larger ones and therefore move faster. The *radioactive* oligonucleotides move with the different non-radioactive fronts and are fractionated according to their affinity for the DEAE groups. The principle of the method termed 'homochromato-

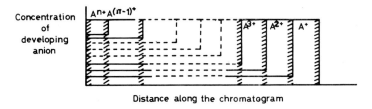

Distance along the chromatogram

Fig. 5.8. Diagram illustrating 'homochromatography'. A mixture of radioactive nucleotides is applied at one end of a DEAE-paper chromatogram (left-hand end of figure), which is developed with a concentrated mixture of anions (non-radioactive nucleotides $A^+, A^{2+}, A^{3+} \ldots A^{(n-1)+}, A^{n+}$). The anions saturate the DEAE-paper and displace one another to form a series of fronts on which the non-radioactive nucleotides (indicated by the shaded areas) are carried.

graphy' is illustrated in fig. 5.8. A suitable 'homomixture' can be prepared from a degraded preparation of yeast nucleic acid (appendix 2) in 7 M urea at a defined pH. If used at high concentrations (e.g. 5%) it should be *dialysed* to remove any excess salt present which otherwise interferes with the homochromatography. Homomixtures of different composition may also be prepared by alkaline hydrolysis of the crude RNA. Such hydrolysed preparations are more suitable for the fractionation of smaller oligonucleotide fragments whereas the unhydrolysed preparations are more suitable for the larger oligonucleotides.

One problem of 'homochromatography' on DEAE-paper is the low

R_f of the larger oligonucleotides on the system because of their high affinity for the DEAE groups. This restricts the method to oligonucleotides of less than 25 residues in length. However, Brownlee and Sanger (1969) have described an improved method using homochromatography at 60 °C on thin layers of mixed cellulose and DEAE-cellulose which extends the range of fractionation to oligonucleotides of 50 residues in length. By diluting the concentration of DEAE groups, and by using higher temperatures, better fractionations with improved R_f values of larger oligonucleotides were obtained.* In addition material appeared as more compact spots with a consequent improvement in resolution.

5.5.1. A two-dimensional procedure using DEAE-paper

Fractionation is carried out on the first dimension exactly as for the standard two-dimensional procedure by ionophoresis on strips of cellulose acetate at pH 3.5 (§ 4.3). Either the short strips (50 cm) or long ones (85 cm) are used depending on the separation required. As most partial products move relatively fast on this system it is necessary to transfer material between the yellow and blue (appendix 2) dye markers for the second dimension. It is usually (although there are exceptions) not necessary to include material running slower than the blue dye marker. Nucleotides are transferred by the usual blotting procedure (§ 4.3) to a sheet of DEAE-paper (47 × 57 cm) 5 cm from one of the short ends of the paper. After drying, the paper is folded in the middle over a supporting rack and lowered into a chromatography tank, previously saturated with water vapour, and allowed to equilibrate for 2 to 3 hr. 50 ml of the homomixture (see below) is then placed in a trough inside the tank and the chromatogram is allowed to develop by ascending chromatography for 16 to 24 hr at room temperature. The blue dye, used as a marker, has an R_f of 0.4 to 0.5 and the yellow an R_f of about 0.25. The paper is normally dried on the rack in air and a radioautograph prepared. No nucleotides run faster than

* The tin-layer method was developed after the sequence of 5S RNA was known. It is described here in detail, however, because it supercedes the method of DEAE-paper homochromatography used in the work on 5S RNA.

blue dye, whilst most partial products are slower than the yellow. Because of the low R_f values of the larger oligonucleotides, the origin may be placed at 1 to 2 cm instead of at 5 cm from one end of the chromatogram, and developed as above. However, although the larger nucleotides move further off the origin, they are more streaky and the results are probably less useful.

A commercial preparation of yeast ribonucleic acid, dissolved in 7 M urea, was a convenient source of a mixture of oligonucleotides. Thus *homomixture a* was prepared by dissolving 10 g of yeast RNA (appendix 2) in 200 ml of 7 M urea and adjusting the *p*H to 7.5 with 10 N KOH. The final volume was about 205 ml. This mixture gave the separations shown in figs. 5.9 and 5.2. Dialysed and/or hydrolysed *homomixtures* are described under the section dealing with homochromatography on thin layers (§ 5.5.2.3).

After exposing the dried chromatogram to an X-ray film usually for 1-12 hr, spots to be eluted are marked out, numbered and cut out of the paper. In order to remove the urea, the pieces of paper are numbered in pencil and washed, with stirring, in a large volume of 95% ethanol for about 2 hr at room temperature. After allowing the ethanol to evaporate, the nucleotides can then be eluted with 30% triethylamine carbonate, *p*H 10 (§ 4.4). Normally the material is divided into two portions. One half is digested in a capillary tube with approximately 10 μl of T_1-RNase (0.2 mg/ml) in 0.01 M Tris-chloride (*p*H 7.5), 0.01 M EDTA for 4 hr at 37 °C. The other half is digested with pancreatic ribonuclease (0.2 mg/ml) in the same buffer for 2 hr at 37 °C. These digests are analysed by ionophoresis on long sheets (85 cm) of DEAE-paper using the *p*H 1.9 system and running the blue marker half-way. Most of the end products could be *tentatively* identified from their position relative to the products of marker ribonuclease digests of 5S RNA which are always run in parallel (fig. 5.11). To confirm an identification the bands are then eluted and their compositions are determined by alkaline hydrolysis. Alternatively, the T_1-end products may be characterised by further digestion with P-RNase and fractionation by DEAE-cellulose ionophoresis at *p*H 3.5 (§ 4.4.2.1) which is particularly useful if counts are low. This distinguishes between end products

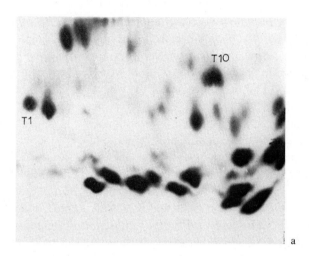

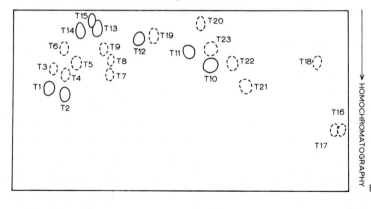

Fig. 5.9. Radioautograph and diagram of a two-dimensional fractionation of a partial T_1-ribonuclease (1:100 for 30 min at 0 °C) digest of 5S RNA using ionophoresis on cellulose acetate at pH 3.5 in 7 M urea in the first dimension, and using *DEAE-paper* and *homomixture a* for homochromatography in the second dimension. Those spots drawn in full lines are visible in the radioautograph whilst those in broken lines are faint or absent and were isolated in other experiments where they were present in better yield. Oligonucleotides moving fast in the second-dimension are end products of T_1-ribonuclease digestion and are not included in the diagram. Analyses are given in table 5.9.

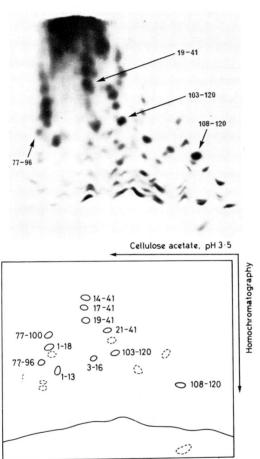

Fig. 5.10. Radioautograph and diagram of a partial T_1-RNase digest (1:4000, 30 min at 0 °C) of 5S RNA fractionated by ionophoresis on a 50-cm strip (Schleicher and Schüll) of cellulose acetate at pH 3.5 in 7 M urea in the first dimension; and homochromatography on a long thin-layer DEAE-cellulose; cellulose (1:10) plate using *homomixture b* in the second dimension. The sequence of some of the prominent spots was determined and is given by residue number of the two ends of the oligonucleotide in the diagram (refer to fig. 5.1 to read off these sequences). To aid comparison of the radioautograph and diagram, some of the sequences shown in the diagram are marked directly on the radioautograph. The wavy line shows the nucleotide front and the single nucleotide ahead of this is G⟩.

having similar mobilities on the DEAE-pH 1.9 system. Some products occupy positions which do not correspond to bands in the marker digest. (CU$_2$) and (C$_2$U) are examples of such products which are

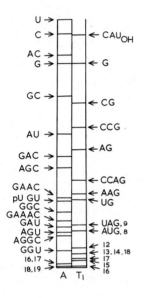

Fig. 5.11. Diagram showing the position of the products of complete ribonuclease T$_1$- and A- (pancreatic) digestions of 5S RNA, fractionated by ionophoresis on DEAE-paper at pH 1.9. The numbered sequences are identified in table 5.7.

generated by T$_1$-RNase digests of some partial P-ribonuclease products.

5.5.2. A two-dimensional procedure using DEAE-cellulose thin layers

This method, suitable for oligonucleotides up to about 50 residues in length, involves ionophoresis on cellulose acetate strips (§ 5.5.1), followed by a transfer of material by blotting onto a thin layer of a mixture of DEAE-cellulose and an excess of cellulose on glass. Fractionation is then carried out at 60 °C in an oven by homochromatography.

5.5.2.1. Preparation of thin layers.
Suppliers of equipment and adsorbants are given in appendices 2 and 3. A slurry of DEAE-cellulose

(MN 300 DEAE), and cellulose (MN 300 cellulose*) is prepared (sufficient for two long glass plates, 20×40 cm, or four short, 20×20 cm, plates) as follows. For the '1 : 10' plates, 1.5 g DEAE-cellulose and 15g cellulose are mixed and added, with stirring, to 102 ml distilled water until the mixture is fairly evenly dispersed and wetted. The slurry is then homogenized thoroughly in a *fast electric blender* for about 3 min to break up aggregated particles. It is then de-aerated thoroughly (10 min on a vacuum pump) and poured into the spreader (giving a fixed 250-μ layer) which is then passed as evenly as possible across the plates positioned on the template. The plates are separated slightly and allowed to dry at room temperature. On inspection they should have no streaks in the direction of spreading. These are usually caused by air bubbles trapped in the slurry thus preventing even flow. If the layer is rather granular this may be due to aggregated particles and homogenisation should be carried out for a longer time. Ridges perpendicular to the direction of spreading are less significant and indicate that the layer was not spread evenly (usually too slowly). Although perfect plates probably give the best separations, quite satisfactory results may be obtained with plates that are somewhat imperfect in any of the above respects. Plates containing a higher proportion of DEAE-cellulose to cellulose, in a ratio of 1 part to 7.5 parts, instead of 1 : 10, may also be prepared. Finally a razor blade is used to define a sharp edge to the thin layers.

5.5.2.2. Fractionation procedure. Partial enzymatic digests are prepared as described below (§ 5.6) and are fractionated in the first dimension, as for the standard two-dimensional system (§ 4.3). Nucleotides are blotted from the cellulose acetate onto the plates as follows. The strip is removed from the tank used for the first dimensional run and monitored with a portable Geiger counter (appendix 3) to find the position of the oligonucleotides, which in partial digests are rather fast-running –

* MN 300 HR is a refined form of MN 300 and may be used as an alternative, in which case 110 ml of water is used. It gives more even plates and better resolution of smaller oligonucleotides (up to 20 residues) than MN 300. It is not known if there is any improvement for larger oligonucleotides (up to 50 residues).

usually in the region just slower than the major red marker. This part of the strip is then placed on the thin-layer plate at 3 cm from one of the short ends and *perpendicular to the direction of spreading of the plate.* Three moist, but not too wet, strips of Whatman 3 MM paper are carefully placed on top followed by a covering glass plate to maintain efficient contact. Water thus flows from the 3 MM paper, transferring the oligonucleotides onto the DEAE-cellulose layer. It is preferable to use strips that still have some 'white spirit' on their surface, rather than let the buffer in the strip dry out and to allow 5–10 min for the transfer. In order to ensure efficient transfer, the paper strips are wetted with water a second time, after carefully removing the covering glass plate. The plate is then repositioned and left for a further 5-10 min. An excess of water or pressure applied to the covering glass plate does not help the transfer but rather tends to dislodge the layer. Transfer is not entirely quantitative and is probably poorer for the larger fragments. However, usually over 80% of the total material is transferred. Because only 20 cm is available for the second dimension, it is convenient to use short runs in the first dimension on approximately 50 cm cellulose acetate strips (1 hr applying 6 kV). This spreads out the partial products less well than the longer runs on 85 cm strips, but allows them to be included in a single second-dimensional run.

Fractionation in the second dimension is carried out at 50-60 °C in an incubator using one or other of the homomixtures described in § 5.5.2.3. The particular mixture to be used depends on whether large fragments, 15 to 40-50 residues long, are to be isolated (homomixture *a* or *b*) or smaller oligonucleotides, 1 to about 20 (homomixture *c*). The thin-layer tank, used for the short plate, or the tall tank (formed by inverting one thin-layer tank on top of another) used for long plates, is equilibrated with about 100 ml of the homomixture at 60 °C in the incubator for 1 hr. The plate is also equilibrated at 60 °C, but outside the tank for the same time. Before starting the ascending homochromatography, it is very important to briefly chromatograph with distilled water until the water front is about 5 cm up the plate. This is designed to wash excess urea, transferred from the first dimension, away from the origin as it otherwise interferes in the fractionation,

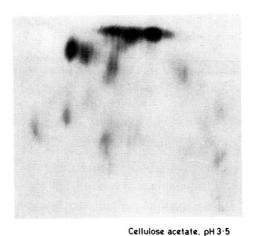

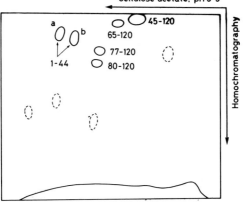

Fig. 5.12. Radioautograph and diagram of a two-dimensional fractionation of a partial T$_1$-ribonuclease digest (1:500, 30 min at 0 °C) of 5S RNA fractionated as in fig. 5.10 except for the following: 'Oxoid' cellulose acetate (57 cm) was used in the first dimension, and homomixture *a* and a short plate in the second dimension. The meaning of symbols is as in the legend to fig. 5.10 and the text. The wavy line at the bottom of the diagram marks the nucleotide front which in this case is close to the solvent front.

especially of faster moving oligonucleotides (as can be seen in fig. 5.12, which was not washed). An alternative procedure, which is quicker and therefore results in less cooling of the plate, is to spray the region of the origin with distilled water. After this, the plate is transferred into the equilibrated tank and ascending homochromatography continued until the front reaches the top of the plate. This usually takes about 2 hr for short plates and 5 hr for long ones, although there is a considerable variation in running time.

The plate is then dried, marked with red ink containing ^{35}S-sulphate, and a radioautograph is prepared by exposing to an X-ray film in a lead-lined folder. It is convenient to prepare two separate exposures (one for an elution template and the other for the records) by exposing two films – one lying on top of the other.

5.5.2.3. Homomixtures. Homomixture *a* is described above (§ 5.5.1). Homomixture *b* is a preparation of homomixture *a* dialysed against 7 M urea for 2-4 hr at 4 °C. In this time there is about a 10% increase in volume but no significant loss of nucleotide material. Homomixture *c* is a dialysed and hydrolysed homomixture prepared by dissolving 10 g of yeast RNA in 100 ml of 1 N KOH and hydrolysing it for 15 min (usually) at room temperature. The solution is then neutralised to *p*H 7.5 with concentrated HCl and is dialysed against distilled water for 2-4 hr. 84 g of urea is then added and the volume made up to 200 ml with distilled water to give a 5% homomixture in 7 M urea. Homomixture *c* may be diluted with 7 M urea to give a 3% mixture, which produces somewhat sharper spots than the 5% mixture, although the R_f values are lower.

Homomixtures *a* and *b* give rather similar separation patterns, although mixture *b* (the dialysed mixture) usually yields more compact spots with better resolution. It is probable that mixture *a* contains some salt (anions) which contribute to the fractionation observed. This contribution may be an advantage for resolving very large fragments, although in general it is a disadvantage, causing some streaking of oligonucleotides which otherwise move as very sharp spots as a result of the homochromatography. Homomixture *c* (hydrolysed and dialysed mixture) separates oligonucleotides from 1 to about 15-20 residues long approximately according to chain length and the main application of this type of homomixture is in the analysis of the longer oligonucleotides in complete T_1-RNase digests of RNA (§ 6.2).

5.5.2.4. Elution and analysis. Those spots to be eluted are marked and numbered on the radioautograph showing the ^{35}S-sulphate ink marks. Holes are then cut with a razor blade corresponding to these spots. This radioautograph then functions as a 'template' and is stuck with transparent tape onto the thin layer, aligning it by means of the ink marks. The thin-layer spots are thus exposed through the holes of this template and may be scraped off, using the elution device shown in

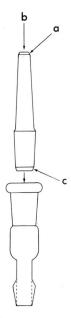

Fig. 5.13. Diagram of a thin-layer elution device with a B10/19 ground-glass joint which can be connected to a vacuum line. The upper part has an orifice (b) of approximately 2 mm internal diameter and is ground on its circumference to give a sharp edge (a) suitable for scraping and dislodging the thin-layer. A sintered disc of porosity 1 or 2 is welded in at (c). Device available from T.W. Wingent (appendix 3).

fig. 5.13, which is connected to a vacuum line. The cellulose is thus sucked up through the narrow orifice (b) and is trapped in the sinter (c). The porosity of the sinter should be number 1 or 2. If it is too coarse, it lets some of the smaller particles through in the succeeding washing and elution procedure, whilst if it is too fine particles do not easily suck up and subsequent elution is very tedious. Ten to twenty samples are usually eluted, using separate top sections of the elution device and the same bottom section. These are then inverted and held in position in a stand with clips. Urea is first removed from the samples by washing with 95% ethanol before the oligonucleotides are eluted with 30% triethylamine carbonate, pH 10. Two or three drops (0.2 ml) are collected into small silicone-treated test tubes, to which a further two or three drops of water is added. The sample is freeze-dried and the triethylamine carbonate is removed in vacuo. Three successive washes with water are used to remove the last traces of triethylamine carbonate.

The partial digestion products are analysed by further digestion with T_1- and pancreatic ribonuclease in the same way as for those partial fragments isolated from DEAE-paper. However, lower enzyme concentrations and shorter digestion times are used because less carrier nucleotide is present on the DEAE-thin layers than on the DEAE-paper. Suitable conditions for both T_1- and pancreatic ribonucleases are as follows. The oligonucleotide is incubated with 10 μl of 0.1 mg/ml of enzyme in 0.01 M Tris-chloride, 0.001 M EDTA, pH 7.4 for about 1 hr at 37 °C.

5.6. Partial enzymic digestion

This is carried out under a variety of conditions depending on the extent of degradation required. The degradation is controlled by varying the ratio of enzyme to substrate and carrying out the reactions at 0 °C. In order to overlap the T_1-end products, it is necessary to study both fairly *extensive* as well as *mild* degrees of partial digestion. The extensive digestion is required so that the smaller partial fragments may be isolated. These are important as it is possible to sequence some of them uniquely from the results of further enzymatic digestion. The milder digestion is required in order to isolate large fragments which overlap the smaller partial fragments. It is only necessary with the very large fragments that their composition expressed in terms of their T_1-end products is known. Both T_1-RNase (§§ 5.6.1 and 5.6.2) and pancreatic RNase (§ 5.6.3) are used to obtain partial fragments using conditions based on Penswick and Holley

(1965), where the reaction is carried out at 0 °C in the presence of 0.01 M magnesium-chloride. The digestion is carried out in 5 μl in the tip of an ice-cold capillary tube and no attempt is made to remove the enzyme before loading directly onto cellulose acetate for fractionation in the first dimension of the two-dimensional system. Digestion with spleen acid RNase is also described below in § 5.6.4.

5.6.1. Extensive digestion with T_1-RNase

A fairly extensive degradation, in which the only *remaining* partial products are derived from *resistant* parts of the molecule, is usually attempted first. About 1 μC of ^{32}P-labelled RNA is digested 30 min at 0 °C with T_1-RNase in a volume of 5 μl in a capillary tube using an enzyme to substrate ratio of 1:100 in 0.01 M Tris-chloride, pH 7.5, 0.01 M magnesium-chloride. Fig. 5.9a shows a fractionation of such a digest using homochromatography on DEAE-paper in the second dimension (§ 5.5.1). The spots near the front are the end products of digestion and the slower-moving and somewhat less sharp spots are the partial products. The spots shown on fig. 5.9a are numbered as shown in fig. 5.9b, which also includes partial products isolated under conditions varying from 1:100 to 1:500 parts of T_1-RNase to substrate. Partial products are analysed by further digestion with T_1- and pancreatic RNase (§ 5.5.1) and the results are collected in table 5.9. In a few cases, indicated by an underlined sequence in the tables, it was possible to deduce the structure of the partial products uniquely from their analyses. In most cases, however, this was not possible and results from other experiments had to be used. For this reason the derivation of the sequences of the nucleotides in this table and others is considered, after all the results have been presented, in § 5.9 below.

A less extensive T_1-RNase digestion (1:4000 and as above) of 5S RNA fractionated using thin-layer homochromatography in the second dimension is shown in fig. 5.10. This fractionation was carried out on a long 1:10 plate using homomixture *b* for the second dimension. Although there are a large number of products they are resolved from one another as discrete spots, except for the faster-moving and smaller products which move as bands, and the slower products where

TABLE 5.9
Analysis of the products of partial ribonuclease T_1 digestion.

Spot no. (fig. 5.9)	Products of further ribonuclease digestion*		Probable structure of nucleotides**
	T_1	A	
T1	UCUCCCCAUG UG G̲	GGGGU GU AU G	UGUGGGGUCUCCCCAUG
T2	UCUCCCCAUG UG G̳	GGGGU AU G	UGGGGUCUCCCCAUG
T3	UCUCCCCAUG UAG U̲G̲ G̲	GGGGU AGU GU AU G	UAGUGUGGGGUCUCCCCAUG
T4	UCUCCCCAUG UAG U̲G̲ C̲G̲ G̲	GGGGU AGU GU AU GC G	UAGUGUGGGGUCUCCCCAUGCG
T5	UCUCCCCAUG UAG U̲G̲ A̲G̲ CG G̲ UCUCCCCAUG	GGGGU G_2,A AGU GU AU GC GGGGU	UAGUGUGGGGUCUCCCCAUGCGAG
T6	UAG AUG U̲G̲ G̲	GGU AGU GU AU G	AUGGUAGUGUGGGGUCUCCCCAUG

(*continued*)

TABLE 5.9 *(continued)*

Spot no. (fig. 5.9)	Products of further ribonuclease digestion*		Probable structure of nucleotides**
	T_1	A	
T7†	pUG CCUG CAG CG G	GGC pU AG GC	pUGCCUGGCGGCAG
T8	pUG CCUG CCG CG G	GGC pU GC G	pUGCCUGGCGGCCG
T9††	CCUUAG pUG CCUG CG G	GGC pU AG GC	pUGCCUGGCGGCCUUAG
T10	UCCCACCUG ACCCCAUG	GAC AU AC G	UCCCACCUGACCCCAUG
T11	UCCCACCUG ACCCCAUG G	GU GAC AU AC G	GUCCCACCUGACCCCAUG
T12	UCCCACCUG ACCCCAUG UG G	GGU GAC AU AC G	UGGUCCCACCUGACCCCAUG

(continued)

TABLE 5.9 (continued)

Spot no. (fig. 5.9)	Products of further ribonuclease digestion*		Probable structure of nucleotides**
	T_1	A	
T13	UCCCACCUG ACCCCAUG UG G̲	GGU GU GAC AU AC G	GUGGUCCCACCUGACCCCAUG
T14	UCCCACCUG ACCCCAUG UG CG G	GGU GAC AU AC G	CGGUGGUCCCACCUGACCCCAUG
T15	UCCCACCUG ACCCCAUG UG CG G̲	GGU GAC AU GC AC G	CGCGGUGGUCCCACCUGACCCCAUG
T16	AACUCAG CCG	GAAC AG	CCGAACUCAG
T17	AACUCAG AAG	A_3,G_2 AAC	AACUCAGAAG
T18	AACUCAG AAG CCG	A_3,G_2 GAAC	CCGAACUCAGAAG
T19	AAACG UAG CCG CG	GU AGC AAAC GC	AAACGCCGUAGCGCCG

(continued)

TABLE 5.9 (continued)

Spot no. (fig. 5.9)	Products of further ribonuclease digestion*		Probable structure of nucleotides**
T20	AACUCAG AAACG UAG UG AAG CCG AG	AGAAGU GAAAC GAAC GU	CCGAACUCAGAAGUGAAACGCCGUAG
T21	AACUG CCAG CAU$_{OH}$ G	AGGC GC AAC AU$_{OH}$	<u>AACUGCCAGGCAU$_{OH}$</u>
T22	AACUG CCAG CAU$_{OH}$ G	AGGC GAAC GC AU$_{OH}$	GAACUGCCAGGCAU$_{OH}$
T23	AACUG CCAG	GGAAC AGGC GC	<u>GGAACUGCCAGGCAU$_{OH}$</u>

* These sequences were identified partly from a knowledge of the position of the end-products of ribonuclease digestion on DEAE-paper ionophoresis at pH 1.9 (fig. 5.11) and partly from the results of further alkaline hydrolysis. Approximate yields of nucleotides were estimated by visual inspection of the radioautographs. If a nucleotide is underlined once there are 2 moles present; if twice there are more than 2 moles. Mononucleotides were frequently run off the end of the paper and are therefore not usually included in the table. In some cases, e.g. T23, CAU$_{OH}$ and AU$_{OH}$ were also run off.

** The structures given are those that can be deduced from the experimental results given in the table, from the T$_1$- and A-end products, and also from a knowledge of other partial products (see text). Where a sequence is underlined it was deduced uniquely from its further degradation products and without reference to other partial products.

† Present in 5 s RNA from strain CA265.

†† Present in 5 s RNA from strain MRE600 only.

the mixture is too complete to resolve. In all cases identical partial digestion fragments had been previously isolated by homochromatography on DEAE-paper (fig. 5.9) so the analysis of these products need not be given in detail. Instead, the sequences are recorded by the residue number (see fig. 5.1 and fold-out in this manual) of the two ends of the fragment in the lower half of the plate. It will be seen that oligonucleotides from 13 to 28 residues long are present as pure products even in this very complicated digest.

5.6.2. 'Half-molecules' with T_1-RNase

An even simpler partial T_1-RNase digestion of 5S RNA fractionated on a short 1 :10 plate using homomixture *a* in the second dimension is shown in fig. 5.12. This digest is designed to cleave the molecule at one or two places only, giving 'half-molecules'. It thus includes most of the labile parts of the molecules, which are cleaved in the more extensive enzymatic digestion described above. This type of digestion can only be achieved if the molecule is isolated in an undegraded state so that particular care (avoiding contaminating nucleases) is needed in the preparation. Once prepared, the RNA is digested and fractionated as soon as possible so as to avoid radiation decomposition. The fingerprint of fig. 5.12 was a result of digesting 5S RNA for 30 min at 0 °C using 1 part of T_1-RNase to 500 parts of substrate in a volume of 5 μl in 0.01 M magnesium-chloride, 0.01 M Tris-chloride, *p*H 7.5. (The reader will notice that these conditions are stronger than those of fig. 5.10, even though the digestion is observed to be less extensive. This may have resulted from using an inactive enzyme preparation, but is more likely to be caused by partial degradation occurring *prior* to the digestion in the experiment shown in fig. 5.10.) The faster moving products are very streaky in this experiment because no attempt was made to chromatograph urea away from the origin before homochromatography (§ 5.5.2.2). The analysis of these products of fig. 5.12 is not given here in detail but the results are summarised in the lower half of the figure by referring to the residue number (see fold-out) of the two ends of the fragment. It will be seen that oligonucleotides of from 41 to 76 residues long are isolated pure in this experiment. It has not been

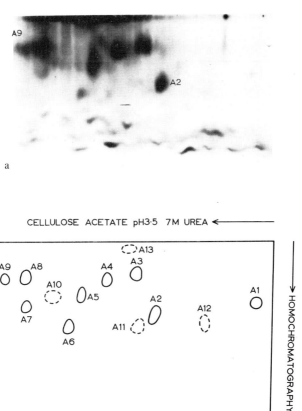

Fig. 5.14. Radioautograph and diagram of a partial pancreatic ribonuclease digest (1 : 500 for 30 min at 0 °C) of 5S RNA fractionated as described in fig. 5.9. The full circles in the diagram are oligonucleotides visible in the radioautograph; the broken circles are the positions of oligonucleotides isolated in other experiments. Analyses are given in table 5.10.

possible to isolate such large fragments on the DEAE-paper and this illustrates a particular advantage of the thin-layer method. Two alternative forms of the sequence of residues 1 to 44 are present, marked 'a' and 'b' on the plate which on analysis are found to differ only in the residue at position 12. This was adenine in spot 'a' and cytosine in spot 'b'. These separate fragments derive from the two major species of 5S RNA (in strain CA 265). They presumably separate in the first dimension on the basis of differences in *shape*, as the single base change would be insufficient to alter mobility on the basis of a compositional difference.

5.6.3. Pancreatic RNase

Fig. 5.14 shows a pancreatic RNase digest of 5S TNA using an enzyme to substrate rate of 1:500 in 0.01 M Tris-chloride, pH 7.5, 0.01 M magnesium-chloride for 30 min at 0 °C. It is fractionated using homochromatography on DEAE-paper in the second dimension.

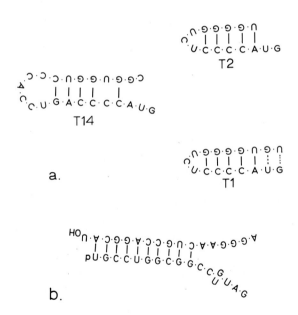

Fig. 5.15. (a) shows a possible base-pairing scheme for some partial T_1-oligonucleotides and (b) a possible base-pairing between the two ends of 5S RNA. A single dash indicates G——C or A——U base-pairs; broken dashes are G– – – –U pairs.

TABLE 5.10
Analysis of products of partial ribonuclease A digestion.

Spot no. (fig. 5.14)	Products of further ribonuclease digestion*		Probable structure of nucleotides[†]
	T_1	A	
A1	$(C_2,A,U)G$ A,C_4	GAC AC	ACCUGACCCC
A2	AUG C_2,A_2,U CCG	GAAC AU GC G	AUGCCGAACUC
A3	AACUG CCAG AG $\overset{=}{G}$ CAU_{OH}	AGGGAAC AGGC GC	$AGGGAACUGCCAGGCAU_{OH}$
A4	AAACG UG AAG AG CCG	AGAAGU GAAAC GU GC	AGAAGUGAAACGCCGU
A5	AUG AG CCG CG G	GGU GAU AGC GC	AGCGCCGAUGGU
A6	AUG $\underline{AG}$ $\overline{CG}$ G	GAGAGU AU GC	AUGCGAGAGU
A7	pUG CCUG CAG CG G	AGU $\underline{GGC}$ $\overline{GU}$ pU GC	pUGCCUGGCGGCAGU

(*continued*)

TABLE 5.10 *(continued)*

Spot no. (fig. 5.14)	Products of further ribonuclease digestion*		Probable structure of nucleotides†
	T_1	A	
A8	pUG CCUG CCG CG G	GGC GU pU GC	pUGCCUGGCGGCCGU
A9	UCUCCCCAUG UG AG CG G	GGGGU GAGAGU AGU GU AU GC	AGUGUGGGGUCUCCCCAUGCGAGAGU
A10	pUG CCUG CG G	pU GGC GC	pUGCCUGGCGGC
A11	AUG CCG G	GGU GAU GC	GCCGAUGGU
A12	AAACG CCG G	GAAAC GU GC	GAAACGCCGU
A13	AACUCAG AAACG AUG AAG UG CCG	AGAAGU GAAAC GAAC GC	AUGCCGAACUCAGAAGUGAAACGC
A14**	pUG CCUG C_2,U	GGC pU GC	pUGCCUGGCGGCCU

(continued)

TABLE 5.10 (continued)

Spot no. (fig. 5.14)	Products of further ribonuclease digestion*		Probable structure of nucleotides†
	T_1	A	
A15**	C,U$_2$ UG AG G	GGGGU AGU GU	AGUGUGGGGUCU
A16**	UG AG U	AGU GU	AGUGU
A17**	C,U$_2$ G	GGGGU	GGGGUCU

* and † See footnotes to table 5.9.
** Isolated by salt gradient chromatography on DEAE-paper.

The products are numbered as shown and the analyses are given in table 5.10.

It may be noted at this stage that fractionation on DEAE-paper by homochromatography is largely according to size. However, there are exceptions. Thus in fig. 5.9 T1 and T2 (17 and 15 residues long, respectively) have roughly the same mobility as each other, and are faster than T10 (17 long). This anomaly may be explained if we assume that T1 and T2 adopt a compact hairpin-looped structure by basepairing and so interact less strongly than T10 with the DEAE-groups on the paper (see fig. 5.15a).

It is apparent that most partial products observed under fairly extensive degradation conditions are related to either T1 or T10 or are derived from the two ends of 5S RNA. This suggests that both ends are somewhat resistant to enzymatic cleavage, and the fact that their sequences are complementary suggests that they pair together to form a stable hydrogen-bonded structure as shown in fig. 5.15b. The products of fairly extensive pancreatic ribonuclease digestion also include resistant fragments from the two ends of the molecule (e.g.

Subject index p. 261

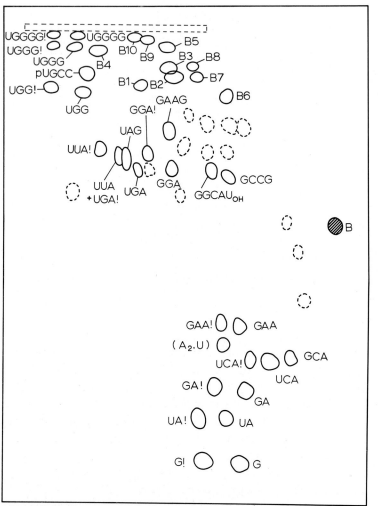

Fig. 5.16. Diagram of a two-dimensional fractionation of a partial acid ribonuclease digest of 5S RNA fractionated by ionophoresis on cellulose acetate at pH 3.5 and on DEAE-paper using 7% formic acid. The area on the origin of the DEAE-paper, included in the dashed lines, was cut out and subjected to homochromatography (fig. 5.17). The analyses of spots B1-B9 are given in table 5.11. For lack of space the customary abbreviation ⟩ has not been used, but in its place a ! indicates an oligonucleotide terminating in a 2′, 3′-cyclic phosphate. B is the blue marker and dashed circles are nucleotides of which the sequence was not determined.

A3 and A8) in good yield. This supports the idea that the specificity of hydrolysis is again dependent on some *secondary structure* feature and is consistent with the scheme in fig. 5.51b. With less extensive degradation fragments other than those already mentioned were isolated, e.g. A4, A5, T16, T17 and probably derive from less stable regions of the molecule.

5.6.4. *Spleen acid RNase*

Fig. 5.16 shows a diagram of a 'fingerprint' of 5S RNA partially digested with acid ribonuclease. Unfortunately, the acid RNase is not commercially available, and was a gift from Dr. G. Bernardi. It was obtained as a 0.2% solution in 0.1 M sodium acetate, pH 5.0. An amount of 5S RNA containing 1 to 10 μC of ^{32}P and 5 to 10 μg RNA is incubated with 2 μl of enzyme for 30 min at 37 °C. The sequences of the oligonucleotides are determined by digesting with T_1 and pancreatic ribonuclease. The results for some of the smaller ones are included in the figure and for the larger ones are given in table 5.11. A large amount of material was still stuck at the origin on the DEAE 7% formic acid run, and in some experiments this was cut out, sewed onto another sheet of DEAE-paper and run on the DEAE-paper homochromatography system (§ 5.5.1). Fig. 5.17 and table 5.12 summarise the combined results from several experiments. Under the conditions used considerable amounts of cyclic phosphates are present. The enzyme appears to have no absolute specificity for any one nucleotide but most splits seem to have occurred after A residues, some after G residues and a few after C residues. The larger products, in particular, provided valuable sequence information (§ 5.9).

5.7. *Chemical blocking with water-soluble carbodiimide*

Gilham (1962) described a water-soluble carbodiimide which formed an adduct with the N^3 and N^1 position of uracil and guanine, respectively (fig. 5.18). The blocked uracil then conferred resistance on the adjacent phosphodiester bond to cleavage with pancreatic RNase, so that a specific cleavage at C residues could be obtained (Lee et al.

TABLE 5.11

Nucleotides from acid ribonuclease digest isolated on the 7% formic acid system.

Spot no. (fig. 5.16)	Products of ribonuclease digestion*		Structure of nucleotide Deduced**	Probable†
	T_1	A		
B1	UG CG A	GA GC U	UGCGA	
B2	UG CCG AA	A_2,G GC U C	UGCCGAA	
B3	UG CCG AAC	$(A_2,G)C$ GC U C	UGCCGAAC	
B4	UG CG G	GGC G U	UGGCG	
B5	UAG CCG CG A	AGC GA GC U C	UAGC(C,G)CGA	UAGCGCCGA
B6††		AAAC GC G C	AAACGCCG	
B7	CG CCG G A	GA $\dfrac{GC}{C}$	GC(C,G)CGA	GCGCCGA
B8	AG CG CCG A	AGC GA GC C	AGC(C,G)CGA	AGCGCCGA

(*continued*)

TABLE 5.11 (continued)

Spot no. (fig. 5.16)	Products of ribonuclease digestion*		Structure of nucleotide	
	T_1	A	Deduced**	Probable[†]
B9	AAACG UG AAAC[‡] CCG G[‡]	GAAAC GC G C U	UG(AAACG,CCG)	UGAAACGCCG

* See footnote * to table 5.9.
** The structures given are those that can be deduced from the experimental results given in the table and from a knowledge of the T_1 and A end products. Inclusion of nucleotides in brackets indicates that their relative order could not be deduced.
[†] These sequences are deduced by making use of results from other partial digestion products and the derivation is described in the text. Where there are many possible structures from the experimental results only the final-deduced sequence is given.
[††] Only detected in one experiment.
[‡] Present in low yield and probably derived by further degradation of AAACG.

1965) without cleavage at U residues. A particular advantage of this reagent is that the blocked group may be removed by mild alkali (pH 11) under conditions which do not cause the cleavage of phosphodiester bonds. This means that an oligonucleotide isolated by the procedure is susceptible to the full range of methods developed for sequencing oligonucleotides. The reagent N-cyclohexyl-N'-(β- morpholinyl-(4)-ethyl) carbodiimide methyl-p-toluene sulphonate (appendix 2), abbreviated CMCT, may thus be used to modify intact 5S RNA so that after digestion with P-RNase some partial products are produced. The chemically modified partial products are conveniently isolated on the standard two-dimensional system. Brownlee et al. (1968) found that it is convenient to attempt partial modification such that only the more accessible U residues are modified. Under the conditions used there is only a small reaction of G residues, which would have considerably complicated the picture had it occurred more extensively. The reaction mixture (10 to 50 μl) contains 1 to 2 μg and

TABLE 5.12 Nucleotides from acid ribonuclease digests isolated by homochromatography.

Spot no. (fig. 5.17)	Products of ribonuclease digestion T_1	A	Probable structure of nucleotide*
B10	UG AG <u>CG</u>	G_3, A_2 GC	UGCGAGAG
B11	UCUCCCCA UCUCCCC** UG G	GGGGU A C U	UGGGGUCUCCCCA +UGGGGUCUCCCC
B12	UCUCCCCA UCUCCCC** <u>UG</u> G	GGGGU GU A C U	UGUGGGGUCUCCCCA +UGUGGGGUCUCCCC
B13	pUG CCUG CG G CC	pU <u>GGC</u> GC C U	pUGCCUGGCGGČC
B14	pUG CCUG CCG CG G	pU <u>GGC</u> GC G C U	pUGCCUGGCGGCCG
B15	AACUG UAG CCAG AG*** G	AGGGAAC AGGC GC C U	UAGGGAACUGCCAGGCA(U_{OH})
B16	UCCCACCUG UAG UG ACCCCA <u>CG</u> G	<u>GGU</u> <u>AGC</u> GAC GC AC	UAGCGCGGUGGUCCCACCUGACCCCA

(*continued*)

TABLE 5.12 *(continued)*

Spot no. (fig. 5.17)	Products of ribonuclease digestion T_1	A	Probable structure of nucleotide*
B17	AACUCAG UG AAG CCG	AGAAG† GAAC GC	UGCCGAACUCAGAAG
B18††	UCUCCCCA UAG UG G	GGGGU AGU GU	UAGUGUGGGGUCUCCCCA
B19†††	UCUCCCCA UAG UG G	GGGGU GGU AGU GU	UGGUAGUGUGGGGUCUCCCCA
B20†††	UCUCCCCAUG UAG UG G	GGGGU GGU AGU GU AU	UGGUAGUGUGGGGUCUCCCCAUG
B21‡	pUG CCUG UAG UG CCG CG G	GGU GGC GU pU AGC GC C U	pUGCCUGGCGGCCGUAGCGCGGUG

* The structures given are deduced from the experimental results shown in the table and from a knowledge of the T_1- and A-end products and of other partial products the derivation of which is described in the text. The sequences underlined could be deduced uniquely from the results in the table and a knowledge of the T_1- and A-end products.
** Present in less than molar amounts.
*** Probably derived from partial breakdown of CCAG.
† Analysed as A_3,G_2. An examination of A-end products shows that AGAAG is the only possible sequence.
†† Contained also traces of UCUCCCAUG, AG and GGU.
††† B19 and B20 were contaminated with each other.
‡ Contained also traces of AAG and several unidentified products.

approximately 1 μC of 5S RNA, between 4.0 and 7.5 mg/ml CMCT (freshly dissolved) in 0.02 M Tris-chloride, 0.002 M EDTA, at pH 8.9. After incubating for 16 hr at 37 °C in a capillary tube the RNA is precipitated by the addition of one-tenth volume of 1.0 M NaCl and

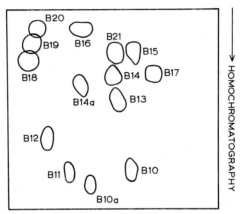

Fig. 5.17. Diagram illustrating a two-dimensional fractionation of a partial acid ribonuclease digest of 5S RNA using ionophoresis on cellulose acetate at pH 3.5 in 7 M urea for the first dimension and homochromatography on DEAE-paper using homomixture *a* for the second dimension. The upper border of the diagram is the origin of the second dimension. In some experiments the material left on the origin after ionophoresis on DEAE-paper using 7% formic acid (see fig. 5.16) was sewed onto another sheet of DEAE-paper and subjected to homochromatography; in others the chromatogram was run directly after the ionophoresis on cellulose acetate. Spots B19, 20, 21 were obtained using mixture *a* at a 10% concentration instead of 5% for homochromatography. The radioactive nucleotides had moved considerably further although the spots were rather 'streaky'. Spots B10a and B14a gave the same sequence as B10 and 14 respectively, and are probably 'cyclic' forms (see table 5.12).

2 volumes ethanol. After recovering the precipitate by centrifugation, the *partially modified RNA* is digested with pancreatic ribonuclease using an enzyme to substrate ratio of 1:20 in 0.01 M Tris-chloride, 0.01 M EDTA, pH 7.5, for 30 min at 37 °C and is fractionated using the

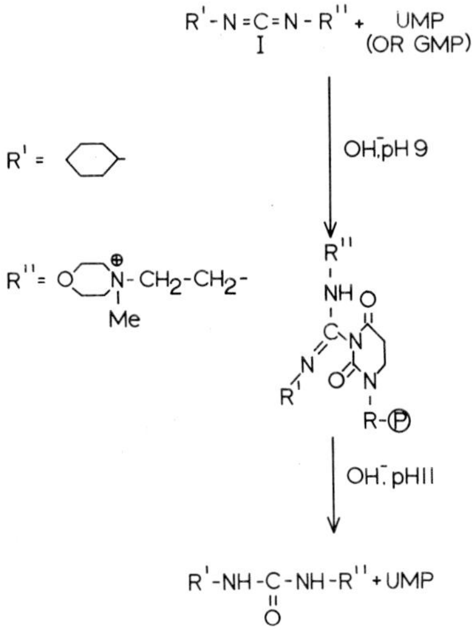

$$R'-N=C=N-R'' + \text{UMP (OR GMP)}$$
$$\text{I}$$

$R' = \langle\rangle-$

$R'' = O\langle\overset{\oplus}{N}-CH_2-CH_2-$
      $|$
      Me

↓ OH⁻, pH 9

[intermediate structure with R'', NH, C, N, R', O, N, R-Ⓟ]

↓ OH⁻, pH 11

$$R'-NH-\underset{\underset{O}{\|}}{C}-NH-R'' + \text{UMP}$$

Fig. 5.18. The reaction of a water-soluble carbodiimide derivative with uridylic (and guanylic) acids.

standard two-dimensional system (§ 4.3). Adequate separation is obtained by fractionation on short lengths (57 cm) of cellulose acetate running the blue marker two thirds of the length of the strip. Oligonucleotides from the cathode end (i.e. origin end) as far as the blue marker are transferred onto DEAE-paper (57 cm long) for fractionation in the second dimension, using the pH 1.9 system. New spots on the 'fingerprint' (not present in a P-RNase digest of a control unmodified 5S RNA) containing a modified U or G, are analysed as follows. After elution the blocking group is removed by incubation with 10 to 20 μl 0.2 N ammonia for about 4 hr at 37 °C. After incubation the solution is dried down on polythene, wetted with water and each spot divided into three parts for analysis (i) by alkaline hydrolysis to establish a composition, (ii) by further digestion with pancreatic ribonuclease, and (iii) by digestion with T_1-ribonuclease. The products of enzymic digestion are fractionated by ionophoresis on DEAE-paper using the

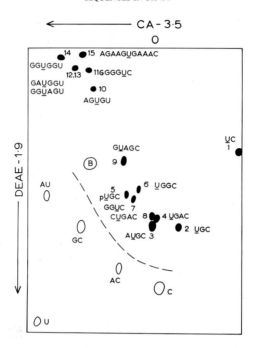

Fig. 5.19. Diagram showing the position of all CMCT-blocked nucleotides containing more than one pyrimidine residue isolated by partial reaction with CMCT followed by pancreatic ribonuclease digestion. Fractionation was by ionophoresis on cellulose acetate, pH 3.5, and DEAE-paper at pH 1.9. The underlining indicates the modified residue. Some unmodified nucleotides (open circles) are included for reference. B is the blue marker, and O marks the origin of the first dimension. The analyses are given in table 5.13.

pH 1.9 system. Marker T_1 and P-RNase digests of 5S RNA are run in parallel with the sample, and the products of digestion are identified by their position relative to these markers and from a knowledge of the composition of the original oligonucleotide.

Fig. 5.19 is a diagram summarising the results of six different digests of CMCT-modified 5S RNA run on the standard two-dimensional system. The 'blocked' oligonucleotides (e.g. GUAGC) lie to the right of the unmodified nucleotides because of their extra positive charge at pH 3.5. Their mobility in the second dimension is also considerably greater than a corresponding unmodified oligonucleotide and is consistent with the addition of two positive charges at pH 1.9. The two charges are on the quaternary ammonium group of the morpholine group and, presumably, on the substituted guanidino-group of the CMCT-base adduct (see fig. 5.18).

The pattern of spots on the two-dimensional system was not entirely reproducible from experiment to experiment, although most spots are consistently present. The strongest spots are C9, C3 and C4, whilst C14 is always prominent. (The symbol C is used to denote a product isolated by the carbodiimide procedure.) C15 is usually weak and C7, C11 and C13 were only detected once. Some modified oligonucleotides with only one pyrimidine residue are present, usually in very low yields. For example, AGGGAAC, GAGAGU and AGAAGU were detected in positions on the 'fingerprint' consistent with the modification of one of their G residues.

Table 5.13 gives the evidence for the sequences of the oligonucleotides isolated as in fig. 5.19. The sequence of many spots was found merely from a knowledge of the P-RNase end-products. Thus C15 gives AGAAGU and GAAAC. Because of the specificity of the blocking reaction for U residues, the sequence must be AGAAGUGAAAC. This was confirmed by the presence of UG in the T_1-RNase digest. Because the blocking of U residues was only partial, some products are obtained which have no C's: for example AGUGU (C10). Among these types of partial products GGUGGU (C14) is present in very high yield, suggesting that its internal U was more reactive than its 3′-terminal U. C8 is unusual in that it has two C's and, although the sequence analysis does not distinguish between CUGAC and UCGAC, the fact that there is no T_1-end product ending in PyUCG (see table 5.7) supports the sequence CUGAC. Pancreatic ribonuclease must thus be hindered in its action adjacent to this 5′-terminal C, perhaps by steric hindrance due to the bulky carbodiimide-blocking group on the adjacent U residue.

From these results, it is clear that it is possible to isolate and sequence fairly easily a specific set of partial products using this method

TABLE 5.13

Nucleotides isolated by ribonuclease A digestion of 5S RNA partially reacted with CMCT.

Spot no. (fig. 5.19)	Composition	Products of further enzymic digest after removal of CMCT*		Structure deduced
		RNase A	RNase T_1	
C1	CU			UC
C2	CGU	GC U	UG C	UGC
C3	CAGU	AU GC	AUG C	AUGC
C4	CAGU	GAC U	UG AC	UGAC
C5	CGpU	pU GC		pUGC
C6	CGU	GGC U	UG G C	UGGC
C7	CG̱U	GGU C	(U,C) G	GGUC
C8		GAC C U	(C,U)G̱ AC	CUGAC
C9	CAGU	GU AGC	UAG G C	GUAGC
C10	AG̱U̱	AGU GU	UG AG U	AGUGU
C11	CG̲U̲	GGGGU C	G̲ (U,C)	GGGGUC
		GGGU**		
		GGU**		
C12	AGU	GGU GAU	AUG G U	GAUGGU
C13†	AG̲U̲	GGU AGU	AUG UAG	GGUAGU
		GAU GU**	G U	(and C12)
C14	G̱U̱	GGU GU**	UG G U	GGUGGU
		G** U**		
C15	CAG̲U̲	AGAAGU	UG AAG	AGAAGUGAAAC
		GAAAC	AAAC AG	
		AAAC** AAC**		
		AC**		
C16		GGC C U		CUGGC

* The nucleotides were identified by their position relative to marker (RNase A and T_1) digests of 5 S RNA (fig. 5.11) on DEAE-ionophoresis at pH 1.9. (UC) lies between G and CG. AAAC runs with AU; and AAC with GC.

** Present in smaller amounts and probably the products of over-digestion with enzyme.

† Contaminated by C12.

which are required in the final derivation (§ 5.9). Although both U and G residues may be modified, the former must react considerably faster (see Gilham 1962) under these conditions. Not all possible residues react equally well which is consistent with a chemical modification occurring more readily in the more open, single-stranded parts of 5S RNA than in the hydrogen-bonded regions (Augusti-Tocco and Brown 1965).

5.8. Chemical blocking by partial methylation

Dimethylsulphate, in aqueous solution, causes the methylation of RNA preferentially at guanine residues giving 7-methylguanine (Lawley and Brookes 1963; Brimacombe et al. 1965). This methylated derivative confers resistance to T_1-RNase and thus suggests a method for obtaining partial products. Partial methylation followed by complete T_1-RNase digestion should therefore allow the isolation of partial products which should have one or more internal m^7G residues, but should terminate in G. The procedure is as follows. In each experiment 5S RNA containing 1 to 10 μC ^{32}P is used. Non-radioactive carrier RNA (appendix 2) is added to give a total of about 20 μg RNA. It is dissolved in 80 μl of 5% sodium acetate adjusted to pH 6.8. 2 μl dimethyl sulphate is added and the mixture is incubated at room temperature with occasional shaking for 20 min. 100 μl water and 20 μl of 20% sodium acetate (pH 5.4) are added and the RNA is precipitated with 450 μl ethanol. After standing at $-20\,^\circ$C for several hours the precipitate is centrifuged, washed with ethanol and dried in vacuo. In order to determine the extent of methylation of the guanosine a small sample is hydrolysed with alkali (§ 4.4.1) and the digest subjected to ionophoresis at pH 3.5. N^7-methylguanylic acid is decomposed by alkali to give predominantly 4-amino-5(N-methyl)-formamido-isocytosine ribotide (Lawley and Wallick 1957), which had $R_u = 0.82$ (R_u = mobility relative to uridine 3'-phosphate) on paper ionophoresis at pH 3.5 (table 7.2). There is in addition an unknown degradation product which has an $R_u = 1.2$, and is present in variable amounts. Under the above reaction conditions about 30 to 50% of the G residues are methylated and this material is suitable for obtaining partial products using T_1-ribonuclease. To the remainder of the methylated RNA is added 1 µl of a solution of T_1-RNase at 1 mg/ml and 2 µl 0.1 M Tris buffer (pH 7.4) containing 0.01 M EDTA (neutralised) and the mixture is incubated at 37 °C for 30 min.

Fractionation is carried out by the standard two-dimensional technique (§ 4.3). A long strip (85 cm) of cellulose acetate is used for the first dimension. After running, it is cut approximately half-way at the position of the blue marker and the two pieces of cellulose acetate are blotted onto two separate sheets of DEAE-paper (85 cm long). That containing material that has moved more slowly than the blue marker on cellulose acetate is run on the pH 1.9 system in the second dimension. The faster moving material may be run on either the pH 1.9 or the 7% formic acid system in the second dimension. Fig. 5.20 shows a typical 'fingerprint' which appears to be quite complex. The sequence of the m^7G containing nucleotides is largely determined by digestion with pancreatic ribonuclease (§ 4.4.2.1) and ionophoresis of the products at pH 3.5 on Whatman No. 540 paper. The positions of the main degradation products in this system are summarised in fig. 5.21. C and m^7GC are not resolved at pH 3.5 so the C band is usually cut out, sewed onto another sheet of paper and subjected to paper ionophoresis at pH 6.5. m^7GC moves at 0.4 times the rate of C at this pH. Several attempts to determine the structure of these m^7G-containing nucleotides by partial digestion with spleen phosphodiesterase (§ 4.6.1.) were unsuccessful. Figs. 5.22 and 5.23, and table 5.14 summarise the combined results from six separate experiments. There is some variation between experiments, probably due to difference in extent of methylation of G and to side reactions. In table 5.14 the sequences deduced are expressed in three different ways. Firstly, the sequences that can be deduced uniquely from the experimental results given in the table are given. The next column makes use of a knowledge of the end products of T_1-RNase digestion (table 5.7). Thus, for instance, M18 [U,(A,G)C]G could be either UAGCG or UGACG or AGCUG or GACUG. However the second, third and fourth possibilities may be excluded as neither ACG nor CUG nor ACUG are T_1-end products.

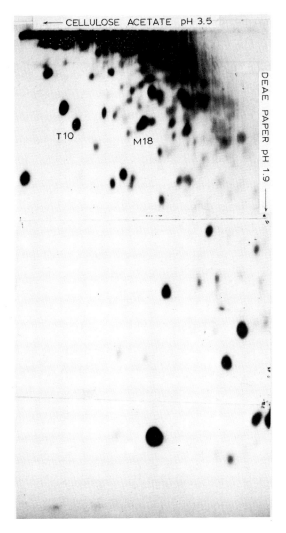

Fig. 5.20. Radioautograph of a two-dimensional separation of a T_1-RNase digest of partially methylated 5S RNA. Fractionation was by ionophoresis on cellulose acetate, pH 3.5 in the first dimension and ionophoresis on DEAE-paper at pH 1.9 in the second dimension. The spots are numbered and identified on the left-hand side of fig. 5.22 and the analyses are given in table 5.14.

TABLE 5.14
Nucleotides from ribonuclease T_1 digests of partially methylated 5S RNA.

Spot no. (figs. 5.22 and 5.23)	Composition*					Products obtained with ribonuclease A	Sequence deduced		
	C	A	U	M**	G		From results	From end products	From other partial products
M1					×	MG	GG		
M1a					×	MG	GG		
M2					×	C MG	CGG		
M2a					×	C MG	CGG		
M3	×			×	×	MC G	GCG		
M4	2.1			0.8	×	C MC G	(C,G)CG	CGCG and/or GCCG***	
M5	× ×			×	×	C MC G	(C,G)CG		
M6	× × ×			×	×	C MC G	(C₂,G)CG		CGCCG
M7	1.0			1.5	1	M̄C MG	GCGG	C(C,G)CG	
M8	2.1			1.7	1	C MC MG	(C,G)CGG	CGCGG or GCCGG	CGCGG
M9		×		×	×	(A,M)G	(A,G)G†	CGAG	
M10	×	×		×	×	C (A,M)G	C(A,G)G	CCAGG or CCGAG	
M11	× ×	×		×	×	C (A,M)G	CC(A,G)G		CCAGG
M12			×	×	×	U MG	UGG	UAGG	
M13		×	×	×	×	MU G	GUG		
M14			×	×	×	MU MG	GUGG		
M15		×	×	×	×	U (A,M)G	U(A,G)G	UAGG or UGAG	UAGG
M16		×	×	×	×	MU AG	GUAG		
M17		1.0	1.2	0.9	1	AU MG	AUGG		
M18	×	×	×	×	×	U (A,M)C G	[U,(A,G)C]G	UAGCG	
M18a		×	×	×	×	U (A,M)C G	[U,(A,G)C]G	UAGCG	

Ch. 5   5S RNA   169

	1.9	1.0	1.0	1			
M19	1.9	1.0	1.0	1	C U MG	(C$_2$,U)GG	CCUGG
M20	× ×	×	×	×	C(A,M)UG	[C$_{1-2}$,(A,G)U]G	(C)CGAUG††	CCGAUG
M21	× ×	×	×	×	C MU AG†††	(C$_2$,GU)AG	CCGUAG
M22		× ×	×	×	A(A,M)U‡ G	A(A,G)UG	AAGUG
M23		× ×	×	×	(A,M)U AG	(A,G)UAG	AGUAG
M24	×(×)	×	× ×	×	C(A,M)C MU G			CCGUAGCG‡‡
M25			× ×	×	U MU G	(G,U)UG	UGUG
M26			× × ×	×	U MU MG	(G,U)UGG	UGUGG
M27					U (A,M) G	[(A,G),U]UG	UAGUG or UGAUG
M28					pU C U MC G	pU(C,U,GC)G	pUGCCUG
M29	×		× × ×	×	U (A$_2$,M)C G	[U,(A$_2$,G)C]G	GAACUG
M30‡‡					C U MC (A,M)C G		UAGCGCCG or UAGCCGCG or AGCGCCUG	UAGCGCCG
M31‡‡					U MU (A,M)U G			UAGUGUG

* Compositions are expressed as the yield of mononucleotide relative to one mole of G. Where the results are given in ×'s, the composition was estimated by inspection of the radioautograph. Where numbers are given in ×'s the composition was determined in the scintillation counter.

** In this table M is 7-methylguanylic acid. It was identified by the presence of its degradation product, 4-amino-5(N-methyl)-formamido-isocytosine ribotide in alkaline digests. Its yield was not quantitative and allowance was made for this in estimating the yield of M.

*** The presence of two spots, giving the same products with ribonuclease A, suggested that both CGCG and GCCG are present. However, this is not conclusive as certain nucleotides (e.g. MG and CMG) appear to be present in two spots.

† If splitting has occurred only at G residues this would have come from the sequence GAGG or GGAG, neither of which is found in ribonuclease A end products.

†† It was uncertain whether this was CGAUG or CCGAUG. The position on the two-dimensional system suggested one C whereas the analysis suggested two.

††† Analysis of the ribonuclease A products on the scintillation counter gave 2.4 moles C to one of MU and AG.

‡ Partial digestion with spleen phosphodiesterase indicated A as the 5'-terminal residue.

‡‡ Various other sequences (e.g. CCGUAGCG, CGUAGCCG) were possible depending on the analysis for C, which was uncertain.

‡‡‡ These nucleotides were detected in only one experiment.

This leaves only the first sequence UAGCG, which is therefore given in the next column of table 5.14. In other cases a unique structure cannot even be deduced in this way, although it is possible to do so if results from other partial digestion products are included. These sequences are given in the final column of table 5.14.

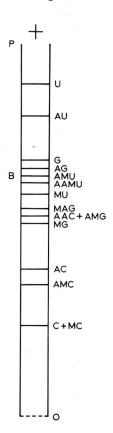

Fig. 5.21. Diagram showing the position of the products of pancreatic RNase digests of nucleotides from T_1-RNase digests of partially methylated 5S RNA fractionated by ionophoresis on Whatman No. 540 paper at pH 3.5. The symbols M, O, B and P refer to m^7G, the origin, the blue marker and pink marker, respectively.

It is apparent from this analysis (figs. 5.22 and 5.23) that the m^7G-containing oligonucleotides are in general rather slower in the first dimension of the fractionation procedure, than the unmodified T_1-end products. This is because of the full positive charge of m^7G below

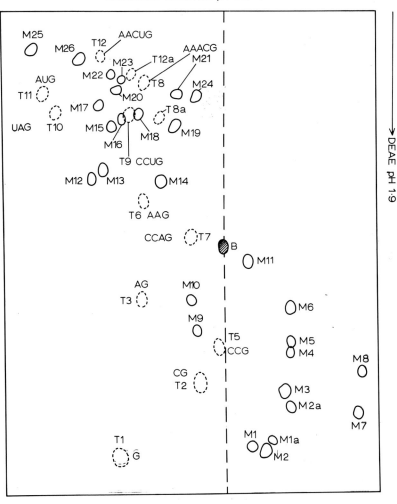

Fig. 5.22. Diagram showing the position of the products of a T_1-RNase of partially methylated 5S RNA fractionated by ionophoresis on cellulose acetate at pH 3.5 in the first dimension. The cellulose acetate was then cut at the blue marker and the two halves (on either side of the dashed line) were fractionated by ionophoresis on two separate sheets of DEAE-paper at pH 1.9 in the second dimension. The analysis of the 'M'-containing (m^7G-containing) spots is given in table 5.14 (see also fig. 5.20). B is the blue marker.

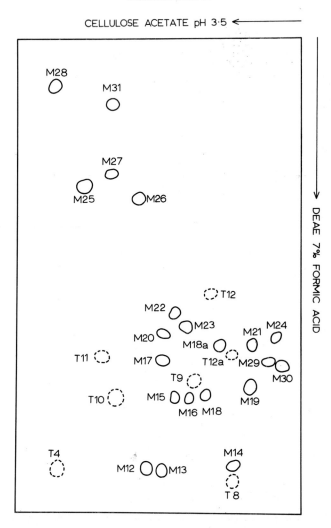

Fig. 5.23. Diagram of a two-dimensional fractionation of a T_1-RNase digest of partially methylated 5S RNA fractionated by ionophoresis on cellulose acetate, pH 3.5 and on DEAE-paper using 7% formic acid. The 'M'-containing spots are identified in table 5.14 and the 'T'-spots (end products of digestion of unmodified 5S RNA) in table 5.7.

pH 7. The two-dimensional procedure is therefore rather suitable for fractionation of these products. Nevertheless the separation of the methylated from unmethylated products was incomplete, especially in the upper half of the fingerprint. Here the separation was usually insufficient to resolve the very large number of products obtained. The complexity undoubtedly reflects the unspecific nature of the methylation (attacking every G residue in the molecule). This contrasts with the partial carbodiimide procedure (§ 5.7). However, another reason for the complexity is the reaction of dimethyl sulphate with A residues giving m^1A (Lawley and Brookes 1963). On alkaline hydrolysis m^1A is converted to m^6A which is indistinguishable from A on ionophoresis at pH 3.5, but which separates on chromatography.

This methylation of A probably accounts for the presence of two spots with the apparent sequence AAACG (T8, T8a), AACUG (T12, T12a) and UAm^{7}GCG (M18, M18a). It would not, however, account for two spots with the apparent sequence m^{7}GG (M1, M1a) and Cm^7-GG (M2, M2a). The nature of the side reaction involved here is not clear.

In summary, the complexity of the 'fingerprints' and the difficulty in sequencing the m^{7}G-containing partial products, made this procedure rather more difficult and probably less useful than the carbodiimide-blocking procedure (§ 5.7). Also rather high amounts of ^{22}P-RNA are required. Nevertheless, a large number of products are obtained and are used in the derivation of the sequence described below (§ 5.9).

5.9. Derivation of the sequence by overlapping

The reader is reminded that it may be helpful to refer to the sequence of 5S RNA in fig. 5.1 or to the fold-out at the end of this manual while following the derivation described here. Where reference is made to the residue number of a fragment this refers to the residue number in the finally deduced structure. Also it is *essential if the derivation is to be understood* that the reader refers to tables 5.7-5.14, which give the degradative evidence for a particular fragment, when this fragment is mentioned in the text. The reader will recollect that the symbols t and a

Subject index p. 261

refer to end products of T_1 and P-RNase digestion (table 5.7); T and A are partial products of T_1 and P-RNase digestion (tables 5.9 and 5.10); B refers to acid RNase products (tables 5.11 and 5.12), whilst C and M are carbodiimide and partial methylation products (tables 5.13 and 5.14). It should perhaps be emphasised that in tables 5.9, 5.10 and 5.12, *only the underlined* sequences are those that may be deduced *entirely and solely* from the results given for that particular fragment. Other sequences are derived from a knowledge of *other* partial products.

An outline of the following derivation may be helpful to the reader. A *small fragment*, of known sequence, serves as a starting point for the derivation. For example, the T_1-end product from the 5'-end of the molecule which is pUG (t17 and residues 1-2) may be used. This sequence is then logically extended by overlapping with other partial fragments also containing this unique sequence, until a point is reached where further logical deduction is not possible. This is usually a rather labile region to partial enzymatic cleavage, where there is a lack of fragments covering this region of the molecule. It may also be that there are fragments capable of extending a particular sequence, but that the partial fragments are not *unique* and could equally occur in another part of the molecule. The same procedure is adopted with four other unique fragments, i.e. the 3'-end, CAU_{OH} (t19 and residues 118-120), t15 (residues 25-33), C12 (residues 72-77) and t16 (residues 87-96). These five unique fragments are thus extended such that they account for the *quantitative* results of § 5.4. The final stage is to overlap these five fragments with one another to give a unique solution. This is achieved by using the information present in the very limited partial T_1-RNase digest of 5S RNA where fragments of 45 and 75 residues in length were isolated (fig. 5.12). The specificity of the procedure used to isolate a particular partial fragment is used in order to deduce the residue on the 5'-side of the fragments. Thus, a fragment of a T_1-RNase digestion is preceded by a G. Similarly, oligonucleotides isolated from P-RNase digests are preceded by a pyrimidine residue. Since the products isolated by the partial methylation method are obtained from a T_1-RNase digest, they also should be preceded by a G residue and this appears to be justified except in the case of nucleotide

M9 (A,m^{7}G)G, which was present in small amounts. There is thus some slight doubt with partial methylation in coming to the above conclusion and no sequence is based solely on it.

5.9.1. 5'-terminal sequence (residues 1-16)

pUG is the 5'-terminal sequence. Nucleotide M28 (pU[C,U,GC]G) contained the pU and must therefore start pUG. There are then two sequences possible from the experimental results: pUGCUCG or pUGCCUG. However, the former is impossible as CUCG is not a T_1-end product, so that the 5'-terminal sequence becomes pUGCCUG. pUGC (C5) was found by the partial carbodiimide method and pUG and CCUG were found together in many partial digestion products (T7-T9, A7, A8, A10), all of which also give CG with T_1-RNase and two moles of GGC with P-RNase. Because there are only two GGC sequences in 5S RNA from *E. coli* strain MRE600, both must be in this 5'-terminal sequence. The sequence UGGC (C6) and CUGGC (C16) both of which contain GGC, must thus be included in the above nucleotides. This makes the only possible common sequence pUGCCUGGCGGC (residues 1-11). This is confirmed by M7 (GCGG), which is derived from the sequence GGCGG, since it comes from a T_1 digest. A10 has this common sequence, pUGCCU-GGCGGC, B13 has an additional C, and T8 gives an additional CCG with T_1-RNase, and must be pUGCCUGGCGGCCG (residues 1-13). T9 gives the same degradation products as T8, but instead of CCG has CCUUAG. It must come from the 5'-end of the molecule but its presence cannot be explained on the basis of a unique sequence and it is concluded that there are two different molecules of 5S RNA, one with the sequence pUGCCUGGCGGCCG (1)* and another with the sequence pUGCCUGGCGGCCUUAG (1a). This heterogeneity, which in this case applies only to the strain MRE600, is discussed further below (§ 5.10.1). Nucleotide A14 also confirms (1a) whilst A8 extends (1) by a U. By analogy with sequence (1a),

---

* (1), (1a) and (1b) are used as identification symbols for the sequence preceding the number so as to simplify the following discussion.

sequence (1) may be extended to be pUGCCUGGCGGCCGUAG (residues 1-16), and this is supported by M21 (CCGUAG). Sequence (1a) is not found in other strains studied, including CA265 on which most of the subsequent work was carried out. In this strain, instead, the sequence pUGCCUGGCGGCAG (T7) was present which differed from T8 by having a CAG instead of CCG. A7 extended the sequence T7 by a U residue, giving pUGCCUGGCGGCAGU (1b). Thus only sequence (1) is common to MRE600 and CA265 and for the purposes of the following discussion (1a) and (1b) will be ignored.

5.9.2. 3'-terminal sequence (residues 103-120)

T21 contains the 3'-terminal sequence CAU_{OH} (t19) and analyses uniquely as follows: AACUG is given in a T_1-RNase digest and must come from the sequence GAACUG as it is a T_1-end product (t12). Since AAC and not GAAC is in the P-RNase digest of T21, AACUG must be the 5'-end and CAU_{OH} is the 3'-end. CCAG, from the T_1-RNase digest, and AGGC, from the P-RNase digest, overlap to give CCAGGC. These three sequences account for all the digestion products except GC and must be joined together in the sequence AACUG-$CCAGGCAU_{OH}$ (residues 108-120). A3 gives AACUG with T_1-RNase and AGGGAAC with pancreatic RNase and shows that the two must overlap, which extends the 3'-terminal sequence to AGGG-$AACUGCCAGGCAU_{OH}$ (residues 104-120). B15 gives the degradation products expected from this sequence but has UAG rather than AG in the T_1-RNase digest, thus extending the sequence to UAGGG-$AACUGCCAGGCAU_{OH}$ (residues 103-120).

5.9.3. Residues 76-105

T2 gives UCUCCCCAUG (t16), UG and G with T_1-RNase and GGGGU and AU as the main pancreatic RNase products. C11 has the sequence GGGGUC so that GGGGU must overlap with the 5'-end of t16, and since T2 gives no GU with a P-RNase the UG must join to the 5'-end of GGGGU to give the sequence UGGGGUCU-

CCCCAUG (a* and residues 82-96). A15 contains the GGGGU sequence. It gives (C,U_2) with T_1-RNase which must represent part of the t16 sequence, so that the 3′-end of A15 is UGGGGUCU. AGU must be the 5′-end since AG is obtained in the T_1-RNase digest. (If it were preceded by a G or A, AGU would not be given by P-RNase; if it were preceded by C or U, AG would not be given by T_1-RNase.) A15 also gives GU with pancreatic RNase and two moles of UG with T_1-RNase, and this can only be explained by the sequence AGUGU-GGGGUCU (residues 78-89). A16, A17 and C10 confirm this sequence. This extends sequence (a) above to AGUGUGGGGUCUCCCC-AUG (b and residues 78-96). T1, B11 and B12 agree with and confirm this sequence. T3 and B18 give UAG with T_1-RNase establishing a GU on the 5′-end of (b). T4 differs from T3 by the addition of a CG and GC, thus adding an extra CG on the 3′-end of (b), and T5 gives AG with T_1-RNase and (G_2,A) with pancreatic RNase, adding an AG to the 3′-end to give GUAGUGUGGGGUCUCCCCAUGCGAG (c, and residues 76-100). UGCGA (B1) confirms the 3′-end. UGCGAGAG (B10) extends it by a further AG. AGAG can only come from the P-RNase end-product GAGAGU (a18) giving GUAGUGUGGGG-UCUCCCCAUGCGAGAGU (d, and residues 76-103). This is nicely confirmed by A6, AUGCGAGAGU. a18 is the only sequence ending in GAGU so that M23 (AGUAG), which should be derived from GAGUAG, must overlap with it, extending sequence (d) to GU-AGUGUGGGGUCUCCCCAUGCGAGAGUAG (residues 76-105).

5.9.4. Residues 14-67

Spot T10 gives only two products (t14 and t15) with T_1-RNase and the presence of GAC in the P-RNase digest establishes their order as UCCCACCUGACCCCAUG (e, and residues 25-41). t14 and t15 are also present in a series of progressively larger fragments, T11 to T15. T11 differs from T10 by the addition of a G in the T_1-RNase digest

* A small letter of the alphabet after a sequence is used here and subsequently for identification purposes in the derivation so as to simplify the following discussion. It is distinct from a small letter followed by a number, e.g. **a1** which is an end product of pancreatic RNase digestion.

and a GU in the P-RNase digest. The G must therefore be added to the 5'-end of (e). T12 differs from T11 by an extra UG in the T_1-RNase digest and GGU instead of GU in the P-RNase digest. That GGU is on the 5'-side of sequence (e) may be concluded from B16 which contains GGU and t15 but only the 5' part of t14 (ACCCCA). T12 is therefore UGGUCCCACCUGACCCCAUG (f, and residues 22-41). T13 has an extra GU compared with T12, which therefore adds a G to the 5'-end of (f), and T14 differs from T13 by a CG in the T_1-RNase digest and by having two GGU sequences and no GU in the P-RNase digest. Because there is no GC in the P-RNase digest, the CG must be 5'-terminal and the sequence of T14 is therefore CGGUGGUCCC-ACCUGACCCCAUG (g, and residues 19-41). Other sequences containing GGU are M14 (GGUGG) and C14 (PyGGUGGU), which confirm that the two GGU's are consecutive.

Besides giving two GGU's, t15 and ACCCCA, B16 gives UAG and two CG's with T_1-RNase and AGC and GC with P-RNase. It gives no GU with P-RNase so UAG must be at the 5'-end and must overlap with AGC to give UAGC. To accommodate the two CG's in B16 the sequence of (g) must be extended to give UAGCGCGGU-GGUCCCACCUGACCCCAUG (h, and residues 14-41). This is confirmed by M8 which is derived from GCGCGG. T15 is also consistent with (h).

Spot B2 analyses uniquely for UGCCGAA and B3 for UGCCGAAC. There are two possible T_1-end products starting in AAC that could be involved in the 3'-end of this sequence – AACUG (t12) and AACUCAG (t13). t12 is already involved in the 3'-terminal sequence of 5S RNA (§ 5.9.2 above) so that t13 must overlap with the UGCCGAAC to give UGCCGAACUCAG (residues 40-51). This is confirmed by the isolation of the unique sequence AUGCCGAACUC (A2) and CCGAA-CUCAG (T16) which extend the above sequence to PyAUGCCGAA-CUCAG (i, and residues 39-51). There are only two T_1-end products ending in PyAUG; these are t16 which is already accounted for in residues 87-95 (§ 5.9.3 above) and therefore could not overlap with (i), and t14, which is the 3'-end of sequence (h). Thus (h) and (i) must overlap to give UAGCGCGGUGGUCCCACCUGACCCCAUGCC-

GAACUCAG (j, and residues 14-51). T17 and T18 analysed uniquely as AACUCAGAAG and CCGAACUCAGAAG, respectively, and established an extra AAG on the 3'-end of (i). This was confirmed also by B17 which analysed uniquely as the sequence UGCCGAACUCAGAAG. AGAAG is derived from a17 (AGAAGU) and this is joined to a9 (GAAAC) in nucleotide C15 which is AGAAGUGAAAC. AAAC is present in the T_1-end product AAACG, and B6 and B9 show that AAACG is associated with CCG giving AGAAGUGAAACGCCG (k, and residues 50-64). A12 and A4 allow (k) to be extended by a U, and T20 extends (k) by a UAG. Thus (j) may be extended to UAGCGCGGUGGUCCCACCUGACCCCAUGCCGAACUCAGAAGUGAAACGCCGUAG (residues 14-67).

5.9.5. Residues 65-79

C12 is PyGAUGGU. M16 (GUAG) is derived from GGUAG and overlaps with C12 to give PyGAUGGUAG (l and residues 72-79) since of the three GGU's two are already accounted for in residues 20-25 (§ 5.9.4 above). The fragment A11 gives AUG, CCG and G with T_1-RNase and GGU, GAU and GC with pancreatic ribonuclease and therefore resolves as GCCGAUGGU by overlapping with (l). This extends (l) to GCCGAUGGUAG (m, and residues 69-79). B5, B7 and B8 all terminate at their 3'-end in CGA and the only sequence with which they can overlap is (m). Also A5 must overlap with (m), which can thus be extended to give UAGCGCCGAUGGUAG (residues 65-79).

All the nucleotides present in a complete T_1-RNase digest of 5S RNA are now included in the five sequences deduced and listed in

TABLE 5.15

1. pUGCCUGGCGGCCGUAG
2. GUAGCGCGGUGGUCCCACCUGACCCCAUGCCGAACUCAGAAGUGAAACGCCGUAG
3. (G)UAGCGCCGAUGGUAG
4. GUAGUGUGGGGUCUCCCCAUGCGAGAGUAG
5. GUAGGGAACUGCCAGGCAU$_{OH}$

Subject index p. 261

table 5.15. It will be seen that all the sequences except the first have GUAG at the 5'-end, and all except the fifth sequence have it at the 3'-end. There are thus four GUAG sequences present in the 5S RNA and the remaining problem was to determine which ones overlapped. This may *easily* be done by using the results of a very limited partial T_1-RNase digestion (fig. 5.12). The fragment extending 44 residues into the molecule from the 5'-end clearly overlaps the first two sequences of table 5.15, whilst another fragment 41 residues long including the 3'-end of the molecule (residues 80-120) overlaps sequences 4 and 5 of table 5.15. In addition sequences 3, 4 and 5 are all joined together in a fragment 76 residues long (residues 45-120) in fig. 5.12. Thus, the fragments of table 5.15 may be overlapped to give a unique sequence. Additional confirmation of this is provided by T6 and T19, the former overlapping sequences 3 and 4, the latter overlapping 2 and 3 of table 5.15.

5.10. Discussion

In both strains of *E. coli* that were studied there appear to be two forms discussion of the significance of the sequence is little more than speculative: however, a number of interesting points may be considered.

5.10.1. Sequence variations

In both strains of *E. coli* that were studied there appear to be two forms of 5S RNA, present in about equal amounts. The sequence given in fig. 5.1 is common to both strains. Strain MRE600 contains a second form of 5S RNA in which the G at position 13 is changed to a U whilst strain CA265 contains a second form in which the C in position 12 is changed to an A. Since a batch of *E. coli* grown from a single colony contained *both* forms, it may be deduced that the two forms are under the control of two separate genes within the single organism. From hybridization studies Morell et al. (1967) suggest that there are a number of genes controlling the production of 5S RNA, so that it is perhaps not surprising to find two forms. Indeed one might have expected to find many more differences and it seems surprising that it was possible to determine a unique sequence. Preliminary studies on some of the minor components in T_1-RNase digests of 5S RNA suggest that there may be other forms of 5S RNA with one or two base changes present in smaller amounts (10 to 20%). As there is 1 molecule of 5S RNA per ribosome it seems likely that 5S RNA molecules containing these variations in sequence are functional.

5.10.2. Secondary structure

Certain regions of the sequence were particularly resistant to digestion with ribonucleases and this was presumed to be due to their being involved in hydrogen-bonded, double-stranded structure (fig. 5.15). Results by other workers on transfer RNA molecules have shown that this is the most likely explanation for resistance, with the various loops being formed by suitable base-paired sequences. This conclusion is strengthened for 5S RNA by the finding that the sequences in the resistant regions are such as to allow plausible, stable base-paired structures. However, it is not possible, even with this information to deduce a unique base-paired scheme for the molecule for a number of reasons. Firstly, the amount of helical structure was unknown and, secondly, the rules regarding the stability of possible alternative base-paired forms – especially of the shorter helical regions – were unclear. Thirdly, no attempt was made to ensure that a single physico-chemical form of the molecule was present. In fact, it seems likely that many preparations of 5S RNA isolated by acrylamide gel electrophoresis are mixtures of at least two forms which are resolved into two bands in long runs (appendix fig. A3). These two forms could be the native and denatured forms which have different mobilities on several column chromatographic systems (Aubert et al. 1968). One might, therefore, be wary of even *expecting* a unique solution for the secondary structure, from the results of partial digestion of the ^{32}P-labelled material.

Nevertheless, it is tempting to speculate about a secondary structure model, solely from sequence studies and Brownlee et al. (1967, 1968) have suggested two forms (fig. 5.24) which they regarded as *minimum* base-paired structures. These had 23 base pairs, including 3 GU pairs. Model 'a' of fig. 5.24 incorporated the results from the partial digests shown in fig. 5.15 whilst model 'b' was an alternative form in which it was assumed that the looped structures T1 and T14 of fig. 5.15 were formed *after* digestion and were *not* a feature of the intact molecule. Physico-chemical studies on 5S RNA showed (e.g. Cantor 1967) that

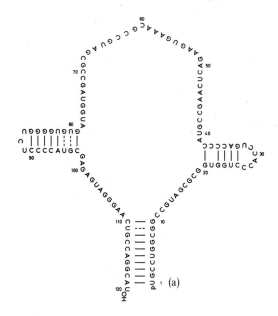

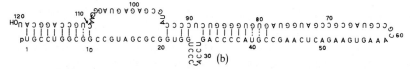

Fig. 5.24. The secondary structure of 5S RNA of *E. coli*. (a) and (b) are two possible *minimum* base-paired structures (Brownlee et al. 1967, 1968). A solid line indicates a standard (G—C or A—U) pair and a dashed line a G---U pair, for example between residues 81 and 95 in (a).

the molecule had a higher degree of helical structure than proposed in these models. Boedtker and Kelling (1967) have estimated a helical content of about 63% by several different physico-chemical methods.

In addition to this, there is evidence that 5S RNA is more asymmetric than tRNA from sedimentation studies. This led these authors to propose a rather elongated and more extensively base-paired structure (fig. 5.25) which, however, was essentially based on model 'b'. It has several features to commend it. Firstly the G residues in the various 'looped out' regions are consistent with the regions observed to be

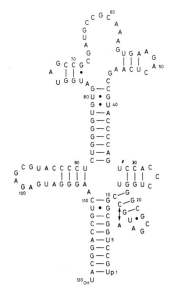

Fig. 5.25. Secondary structure of 5S RNA. (From Boedtker and Kelling 1967.)

rather labile to partial T_1-RNase digestion. Secondly U at positions 22 and 32 are not paired explaining their availability to the carbodiimide reagent (sequences C4 and C14 of table 5.13). Unfortunately the other sequences reacting with this reagent (C3 and C9 of table 5.13) both occur twice in the molecule and it is only possible to say that at least one of each of the U's in these sequences must be available to modification by the reagent. This is in fact the case in the model where U(95) and U(65) are free. However, against the model are two points. Firstly Lee and Ingram (1969) have deduced, from reaction with carbodiimide, that U(40) is available, whereas this region is base-paired in the model. Secondly the various alternative sequences that have been detected A(12) and U(13) would be expected to decrease the stability of the short helical region in which they occur as they both interfere with a base pair. This would be unexpected if this region were important for the structure (or function) of the molecule.

In conclusion it may be said that any model of 5S RNA must be regarded as tentative until a definitive attempt at X-ray crystallographic analysis of a single physico-chemical form is made. In the case of the model discussed here, the arrangement of the shorter helical regions (of 4 base pairs or less) is less certain than that of the two longer, helical regions of 10 base pairs each, which are more likely to be correct. Without model building it is difficult to envisage a three-dimensional structure, although it would appear from fig. 5.25 that two of the shorter helices could be extensions of the two longer helices without interruption of base-pairing. This would, presumably, be a further argument in favour of the arrangement shown in the model as helices of 14 base pairs each would possess a considerable degree of stability. It may be noted that, even with this model, not all possible base pairs are made and it is conceivable that there is more helix than the physico-chemical studies indicate. A good example is a possible pairing of residues 94 and 95, with 103 and 102, respectively. This region and the region between residues 43 and 78 may be the parts of the molecule denaturing most easily and may be regions of lower orders of stability than the extended helical regions already mentioned.

Perhaps a reasonable approach in trying to deduce a tentative secondary structure for any newly established sequence (in the absence of any physico-chemical data) is to attempt to maximise base pairing giving particular attention to long accurately paired regions, irrespective of whether they occur in nearby or distant parts of the molecule. If hair-pin loops are postulated they should have a minimum of three bases in the loop. The results of partial digestion on the molecule should be used with caution, and the fact that a hair-pin loop is isolated as a looped structure does not necessarily mean that this loop exists as such in the intact molecule. Of course, if the new structure is a tRNA then it should conform to the standard clover-leaf arrangement.

Any proposed structure can be further tested against information from reaction with any chemical reagent (e.g. a carbodiimide) reacting under conditions where the secondary structure should still be intact (i.e. near neutral pH and physiological temperatures).

5.10.3. Sequence homologies

There are two sequences in 5S RNA of ten and eight residues, respectively, that are repeated in the molecule. In fig. 5.26 the structure is written so that the common sequences are aligned. It is apparent from this type of analysis that in addition to these two regions there is considerable *further* homology between the two halves of the chain.

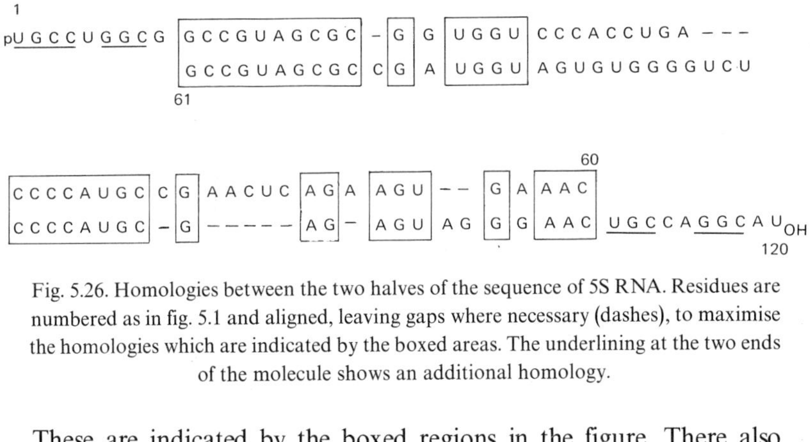

```
 1
pU G C C U G G C G  | G C C G U A G C G C | - | G | G | U G G U | C C C A C C U G A - - -
                   | G C C G U A G C G C | C | G | A | U G G U | A G U G U G G G G U C U
                    61


                                                                     60
| C C C C A U G C | C | G | A A C U C | A G | A | A G U | - - | G | A | A A C |
| C C C C A U G C | - | G | - - - - - | A G | - | A G U | A G | G | G | A A C | U G C C A G G C A U_OH
                                                                               120
```

Fig. 5.26. Homologies between the two halves of the sequence of 5S RNA. Residues are numbered as in fig. 5.1 and aligned, leaving gaps where necessary (dashes), to maximise the homologies which are indicated by the boxed areas. The underlining at the two ends of the molecule shows an additional homology.

These are indicated by the boxed regions in the figure. There also appears to be some homology between the two ends of the molecule as shown by the underlining in the figure. One interpretation of this observation is that 5S RNA evolved from a smaller RNA, of about half the size of present-day 5S RNA, by a duplication of the gene (DNA) for 5S RNA. A rather similar explanation has been suggested by Fellner and Sanger (1968) to explain the occurrence of duplicated methylated sequences in 23S RNA. However one must also seek some explanation for the fact that homology is shown in certain regions but that between them there is no homology. Presumably the preservation of the homologous sequences against mutations is best explained if they are

Subject index p. 261

essential for the structure and function of the molecule. From this one might speculate on a symmetrical function for the molecule; i.e. it could have two similar active centres or binding sites, although this is not apparent from the secondary structure models. On this basis the possibility cannot be excluded that the homologies may have arisen by convergent evolution from a single gene rather than by divergent evolution from a duplicated structure.

Forget and Weissman (1967) have recently determined the sequence of the 5S RNA from human carcinoma (KB) cells. Although it shows a certain amount of homology with the *E. coli* material it is surprisingly different and does not show many of the features discussed above. The main common features are that both 5S RNA's are 120 residues long, both show base-pairing between the two ends, and in both an elongated model may be constructed. Other possible base-pairing residues seem to differ in the two molecules. The KB cell material contains repeating sequences but different ones and in different positions from those found in *E. coli* material. If we assume that there was a common ancestral 5S RNA we can only conclude that there has been a considerable divergence in its separate evolution in bacteria and Man.

CHAPTER 6

Methods for sequencing large oligonucleotides isolated as end products of T_1-RNase digestion

6.1. Introduction

The method of partial digestion with venom phosphodiesterase is undoubtedly the most useful general procedure for sequencing oligonucleotides in the size range of tetra- to decanucleotides which are derived by complete T_1-RNase digestion of an RNA. This is amply illustrated by its use for sequencing, oligonucleotides from 5S RNA (§ 5.2.2). However, larger oligonucleotides than decanucleotides may be expected on a statistical basis and have, in fact, been isolated as end-products of T_1-RNase digestion from molecules other than 5S RNA. The sequence of these large end products usually presents some difficulty, principally because the general method of sequencing by partial digestion with venom phosphodiesterase often proves to be unsuccessful. One obvious difficulty with the longer oligonucleotides is their greater susceptibility to secondary splitting (§ 5.2.1) or to an endonuclease contaminating the exonuclease activity of the venom phosphodiesterase. Thus an oligonucleotide may cleave into two halves before the exonuclease activity of the enzyme has time to work. Another difficulty with the larger oligonucleotides is that they might have a sequence which allows them to adopt a hair-pin looped structure by forming Watson-Crick base pairs. As double-helical RNA is resistant to digestion with venom phosphodiesterase it is necessary to denature the oligonucleotide by heating and rapid cooling prior to adding the enzyme (Min-Jou and Fiers 1969). There remains, in addition, the prob-

Subject index p. 261

lem of the fractionation of all the intermediate products. Thus, even supposing that all possible intermediates of a partial digest of an oligonucleotide are present in sufficient yield, these may not be separated by one-dimensional ionophoresis on DEAE-paper at pH 3.5 (§ 5.2.2). On this fractionation system intermediates larger than hepta- or octanucleotides will not separate from one another and remain at the origin. Even using ionophoresis on DEAE-paper and the 7% formic acid system, products with four or more U residues have an insignificant mobility and may not separate from one another. In fact it may be necessary to use a two-dimensional analysis using the standard method described in ch. 4 to separate all the products of a digestion (e.g. see Dahlberg 1968). With very large oligonucleotides even this may fail and it may be necessary to use homochromatography on thin layers of DEAE or to use the two-dimensional system using homochromatography in the second dimension – a method described in § 5.5. However, it should be emphasised that the major problem is not the fractionation of the intermediates of a partial venom phosphodiesterase digest, but the fact that the intermediates may be absent or present in insufficient radioactive yield to obtain an analysis of them.

Although it is difficult to give dogmatic advice it would appear that for an oligonucleotide of ten or more residues in length and of average composition (e.g. C_3, A_3, U_3, G) it is preferable to adopt an alternative procedure to partial venom phosphodiesterase digestion. This involves isolating the end products of digestion after cleaving the oligonucleotide *base-specifically*. Thus cleavage at A residues may be achieved by digestion with RNase U_2, and at C residues by using P-RNase after chemical blocking of the oligonucleotide by means of a water-soluble carbodiimide (CMCT). The sequence is then reconstructed by overlapping these products. An example of this approach is given below in § 6.3, and allows the sequence of a 21-residue long nucleotide to be established. However, in other cases these two base-specific procedures may give insufficient information to allow the sequence to be deduced and it may be necessary to degrade the oligonucleotide partially. Such was the case with an oligonucleotide 17 residues long isolated from 6S RNA of *E. coli* and in this case *partial* digestion with spleen acid RNase, which cleaves preferentially, but not exclusively, at A residues, was used. This enzyme seems particularly useful for achieving limited cleavage of oligonucleotides but as it is not yet commercially available it should be mentioned that partial digestion with pancreatic RNase may also be used in order to overlap the products of the base-specific cleavage. The deduction of the sequence of the 17-residue long oligonucleotide from 6S RNA is also discussed below (§ 6.4) after a short discussion of the problem of isolation of large oligonucleotides (§ 6.2). The experimental details of the sequence methods used appear at the end of this chapter (§ 6.5-6.8).

It is still true that for large oligonucleotides *with few or no A residues* (e.g. C_6, U_6, G) which are resistant to RNase U_2, it is probably best to use partial digestion with venom phosphodiesterase as a sequence method. Such oligonucleotides, lacking an A residue, are neither susceptible to secondary splitting (Py–A cleavage) nor are they likely to adopt a compact secondary structure. They are thus suited to sequence analysis by partial digestion with venom phosphodiesterase (e.g. see Fellner et al. 1970), although it may be an advantage to purify the commercially available enzyme to free it of endonuclease activity and also to use a two-dimensional analysis to separate all the products. The principal method of confirming a sequence deduced by this method would be to isolate the C-terminating products after digestion with P-RNase of the CMCT-blocked oligonucleotides (§ 6.5). Partial digestion with P-RNase on the chemically unmodified oligonucleotide may also give useful results as U-U bonds appear to be more resistant to cleavage than U-C bonds. Thus products such as U_n where $n = 2, 3$, etc. may be isolated in good yield under appropriate conditions (see § 6.8 below).

6.2. *Isolation of large oligonucleotides*

The method of isolation will depend on the complexity of the mixture of oligonucleotides in a digest. In the case where a study is being made of a purified *small molecular weight RNA* of about 100 residues in length, a large oligonucleotide in a complex T_1-RNase digest will usually be encountered as a spot with a low or negligible mobility in the second dimension of the standard two-dimensional ionophoretic system (§ 4.3). It will probably be pure as it is unlikely, on a statistical basis, that there will be present more than one large oligonucleotide of identical or nearly identical composition in such a digest. If, however, it is suspected from further analysis that a spot on the origin of the second dimension is impure it may be further fractionated by homochromatography on DEAE-paper (§ 5.5.1) as follows. A strip of paper including the origin spot is cut out and sewed onto a new piece of DEAE-paper (47 × 57 cm). The piece of DEAE-paper lying underneath the strip of paper containing the radioactive oligonucleotide material is cut out and then homochromatography carried out using mixture b or c as described in § 5.5.1.

Where large oligonucleotides are to be isolated from a T_1-RNase digest of a *high molecular weight RNA* it may be possible to use the standard two-dimensional system to isolate large oligonucleotides and obtain most of them pure. However, it will undoubtedly be necessary to isolate such products in a form lacking their 3'-terminal phosphate group by using alkaline phosphatase in combination with T_1-RNase in the preparation of the digest. This is because the larger nucleotides resolve better from one another on the standard two-dimensional system when dephosphorylated (§ 4.5.1). Thus Fellner et al. (1970) have isolated and sequenced most of the T_1-oligonucleotides isolated from 16S RNA of *E. coli* using this method, although in this molecule the largest oligonucleotide was only 14 residues long. Nevertheless, it is probably an advantage when studying oligonucleotides of this size or larger to use the two-dimensional fractionation system which uses homochromatography on thin layers of DEAE-cellulose in the second dimension (§ 5.5.2). This method has exceptional resolution and an example of its application in the isolation of long oligonucleotides from the RNA of the bacteriophage R17, is shown in fig. 6.1. There appears to be a definite advantage in isolating large radioactive oligonucleotides in the presence of the carrier, non-radioactive, oligonucleotides introduced in the homochromatography, as they appear to stabilize the radioactive oligonucleotides against

Ch. 6 SEQUENCING LARGE OLIGONUCLEOTIDES 191

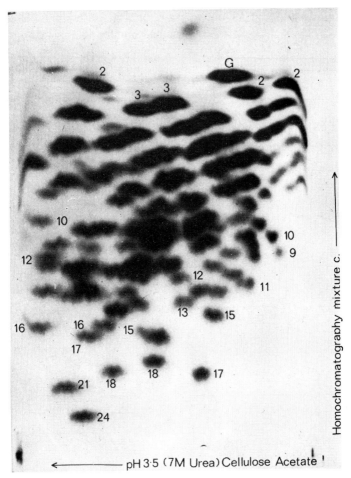

Fig. 6.1. Complete T_1-RNase digest of bacteriophage R17 RNA fractionated by ionophoresis on cellulose acetate at pH 3.5 in the first dimension; and by homochromatography on a 1:7.5 DEAE-cellulose thin-layer plate using *mixture c* in the second dimension. The numbers beside the spots refer to the number of bases in the oligonucleotide.
(Jeppesen 1971)

breakdown. In particular, secondary splitting is not observed. Thus, more accurate and specific cleavage is observed on further treatment of an oligonucleotide with RNase U_2, for example, than when the same digestion is carried out on the same oligonucleotide isolated by means of the standard two-dimensional ionophoretic system.

6.3. Deduction of the sequence of the oligonucleotide AAUUAACUAUUCCAAUUUUCG
(Adams et al. 1969)

Table 6.1. gives the results of four different digests of this fragment labelled 21 in fig. 6.1 from R17 RNA. The composition (column a)

TABLE 6.1
Sequence analysis of oligonucleotide A (from Adams et al. 1969).

Base composition ($G = 1.0$)	Pancreatic ribonuclease digestion products	CMCT blocked pancreatic ribonuclease digestion products	U_2 ribonuclease digestion products
(a)	(b)	(c)	(d)
U 8.6	2 AAU**	C	A
G 1.0	AAC	(AAU*,U*)AAC	AA†
A 6.5	AU	G*	(C,U)A
C 3.9	G	(AAU*,U$_3^*$)C	UUA
	3 C**	(AU*,U$_2^*$)C	UUAA†
	6 U**		$(U_2,C_2)A$
			$(U_2,C_2)AA$†
			$(U_4,C)G$

* The asterisk indicates the residue (U or G) blocked by the carbodiimide reagent.
** Yield estimated by visual inspection of the radioautograph.
† Partial digestion products.

was determined quantitatively but gave ambiguous results with respect to the number of U and A residues. However, even a visual analysis of the results of a P-RNase digestion (column b) indicated that there were seven A residues accounted for by the following products: two AAU's, an AAC and an AU residue. It should thus be clear to the reader that it might have been better to quantitate the results of digest (b) rather than (a) in this particular experiment. The results of digest (c) gives the composition of the four products terminating in C residues, in addition to the mononucleotide G. These CMCT-blocked products were analysed by further digestion with P-RNase (after removing the blocking group with ammonia (§ 6.5)) and together account for all the P-RNase products already characterised in column (b) of table 6.1. The results in column (d) giving products ending in A are somewhat confusing as the results include some *partial* digestion products terminating in AA. These partial products are often present in better yield than the A-ending products, and their presence is particularly useful in establishing overlaps between products in the separate digests. Thus the sequence may be reconstructed as follows. The presence of free G in digest (c) indicates, from the specificity of cleavage, that G is preceded by a C residue. This allows the nucleotide $(U_4,C)G$ in digest (d) to be resolved as UUUUCG. As this nucleotide is present in a U_2-RNase digest (d) we may deduce from the specificity of the enzyme that the above sequence will be preceded by an A residue giving AUUUUCG. This product must then overlap with the product $(AAU,U_3)C$ in digest (c) extending the derivation to AAUUUU-CG. (The other products in digest (c) may be excluded as they have fewer than four U residues.) AA in digest (d) can only derive from the 5'-end of the whole oligonucleotide. (If it were internal it would be connected to nucleotides other than A as is UUAA, for example.) AA could overlap with either of the AAU containing products of digest (c). However, as one of the possibilities has already been placed, by overlapping near the 3'-end, this leaves the other product (AAU,U)AAC, to overlap with the 5'-terminal AA to give AAUUAAC. This sequence is confirmed by the presence of UUAA in digest (d). This leaves only one other AA-ending product in digest (d) not already accounted for, that is $(U_2,C_2)AA$ which must overlap with the sequence AAUUUU-CG (already deduced) giving $(U_2,C_2)AAUUUUCG$. The sequence within brackets may be resolved by considering the remaining two products – C and $(AU,U_2)C$ in digest (c).

Theoretically, these could be present as either $C(AU,U_2)C$ or as $(AU,U_2)CC$. However, only the second possibility, which places the two C residues together, can overlap with $(U_2,C_2)AAUUUUCG$ to give UAUUCCAAUUUUCG. (The first possibility places three U's between the two C residues which is incompatible with (U_2,C_2)-

AAUUUUCG.) This may be joined to the 5'-terminal sequence already deduced (AAUUAAC) to give the complete sequence, AAUU-AACUAUUCCAAUUUUCG. The remaining product (C,U)A of digest (d) confirms the overlap between them. Fig. 6.2a shows the position of the products in (c) and (d) of table 6.1 in the final sequence that was deduced.

It perhaps should be emphasised that for this particular oligonucleotide, the presence of several strategically placed AA sequences, combined with the presence of partial products terminating in AA in the U_2-RNase digest, made this derivation possible without resorting to partial acid RNase or partial P-RNase digestion. Such a procedure is usually required as exemplified below in § 6.4.

6.4. Deduction of the sequence of the oligonucleotide AUAUUUCAUACCACAAG
(Brownlee 1971)

Table 6.2 gives the results of five different digests of this oligonucleotide isolated from 6S RNA of *E. coli*. The reader will notice a discrepancy in the total number of U residues in column (a) which is 5, and under column (b) which totals 6 – made up of three U's as a mononucleotide and three other U residues included in the dinucleotide sequence, AU. Digest (c), (d) and (e) confirm that there are only five U's present and that there is a mistake in column (b). This illustrates the advantage of

(a)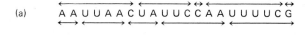

(b)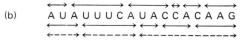

Fig. 6.2. Schematic description of derivation of the sequence of (a) an oligonucleotide from bacteriophage R17 RNA; (b) an oligonucleotide from 6S RNA of *E. coli*. A ↔ *above* the sequence indicates a product derived by P-RNase of the CMCT-blocked oligonucleotide, the same symbol *below* the sequence is a product of U_2-RNase digestion. The symbol ←--→ indicates a product of spleen acid RNase digestion.

TABLE 6.2
Sequence analyses of an oligonucleotide from 6S RNA (Brownlee 1971).

Composition (a)	P-RNase digestion (b)		CMCT and P-RNase digestion (c)	U_2-RNase digestion (d)	Spleen acid RNase digestion (Main products) (e)
C 4.4	C	2.3	(AU*,U$_2^*$)C	UUUCA	AUA
A 6.7	AC	1.9	AU*AC	CAA†	[U,(AC)$_2$,C]AAG
G 1.0	AU	2.7	AAG*	CA	(U$_3$,C)A
U 5.1				2 UA**	
	U	3.0	AC	CCA	
	AAG	1**	AU	A & G	
			C		

* The asterisk indicates the residue (U or G) blocked by the carbodiimide reagent.
** Yield estimated by visual inspection of the radioautograph.
† Partial product.

performing more sequence techniques than the minimum required to sequence an oligonucleotide, as thus mistakes are easily detected.

Notice also that, unlike the results in table 6.1, digest (c) includes the product AU – a product ending in U, in addition to products terminating in C and G. This results from incomplete modification of the oligonucleotide with the carbodiimide-blocking reagent, thus allowing a limited cleavage at U residues, in this case adjacent to AU. Digest (d) gives the A-terminating products including, as in table 6.1, a partial product – CAA. The sequence UUUCA in (d) was established by the determination of its composition, its 5′-end group (§ 4.4.2.4) which was U, and by the isolation of the fragment UUU in a partial pancreatic digest (§ 6.8).

From the results of digests (c) and (d) the following may be deduced. UUUCA may be overlapped with (AU,U$_2$)C giving *AUUUCA*. In addition the partial product CAA in digest (d) may be overlapped with AAG in digest (c) giving CAAG. The specificity of U$_2$-RNase allows this to be extended by an A at the 5′-end giving *ACAAG*. These two fragments cannot be further extended logically without a knowledge of the partial products in digest (e). These three partial acid

RNase products were partially sequenced by further digestion with pancreatic and U_2-RNase giving the information shown in the Table 6.2. They can *only* be arranged in the order AUA−UUUCA−(U,-(AC)$_2$,C)AAG as the alternative order UUUCA−AUA−(U,(AC)$_2$,-C)AAG would produce an AAU, which is not the case (see digest b). From our previous overlapping this resolves the sequence to AUAU-UUCA[U,(AC)$_2$,C]AAG. The presence of AUAC in digest (c) further resolves this sequence to AUAUUUCAUAC(AC,C)AAG. This may be overlapped with ACAAG already deduced to give the sequence AUAUUUCAUACCACAAG, which is in agreement with all the other results in table 6.2. Fig. 6.2b shows the position of the products in digest (c), (d) and (e) of table 6.2 in the final structure.

In summary a general procedure for long T_1-oligonucleotides is to characterise the products of (1) pancreatic RNase digestion (2) pancreatic RNase digestion after carbodiimide blocking and (3) U_2-RNase digestion. Should the required information for establishing a sequence be incomplete, then a further partial digest with spleen acid RNase may be attempted.

6.5. *Cleavage next to C's*

The ability of a water-soluble carbodiimide to react with uridine residues in 5S RNA and thus block the action of pancreatic RNase has already been discussed in § 5.7. The reagent may also be used to obtain specific cleavage of T_1-end products at C residues. The method is applied to T_1-end products isolated by using the standard two-dimensional system (§ 4.3). In the experiments described in this chapter the reagent CMCT (appendix 2) was made up freshly in 0.01 M Tris-chloride, 0.001 M EDTA, *p*H 8.9 at a concentration of 20-50 mg/ml. Approximately 10 μl was used for incubation in sealed capillary tubes overnight at 37 °C. About 10 μl of 0.2 mg/ml pancreatic RNase in 0.1 M Tris-chloride, 0.01 M EDTA, *p*H 7.4 was added to make a final volume of approximately 20 μl and the incubation continued for a further half to one hour at 37 °C. Very recently, Dr. U. Rensing has found that the reaction with CMCT is better at *p*H 7.5 (in 0.05 M Tris-chloride, 0.005 M EDTA) than at *p*H 8.9. After reaction (see above) 10 μl of P-RNase (0.2 mg/ml in water) is added and the incubation continued as before. The products are fractionated by ionophoresis at pH 3.5 on Whatman No. 540 or No. 3 MM paper (47 × 57 cm) with the origin placed 30 cm from the negative end of the paper for 1-2 hr at 2-3 kV. Oligonucleotides that are chemically blocked by the reagent are neutral or move towards the cathode because of the two additional positive charges introduced by substitution with the blocking group at pH 3.5. When applied to oligonucleotides isolated by means of '*homochromatography*' where an excess of carrier nucleotides is present, a more concentrated solution of CMCT is necessary and a solution at 100 mg/ml must be used. Also the subsequent P-RNase digestion is carried out for 2 hr at 37 °C. Fractionation is on Whatman No. 3 MM paper (which has a higher capacity than No. 540 paper) because of the high concentration of reagent and carrier nucleotides. The position of some of the commoner CMCT-blocked oligonucleotides on this system is shown in fig. 6.3. It is difficult to rely on the exact position of bands as the high concentration of the reagent does influence the mobility of nucleotides. Normally, however, the smaller products move well as sharp bands, (e.g. U*C and (AU*)C) whereas the larger products show more tailing, (e.g. $(A_2U_2^*)C$). Another difficulty is that incompletely modified products may be present and confuse a tentative identification based on mobility. An example of a partially modified product is (AU*U)C in fig. 6.3. In addition, some partial products of the pancreatic RNase digest may also be present, and products such as (CU*)C and $(CU_2^*)C$ have been observed even after extensive digestion with the enzyme. Presumably they are resistant because the blocking group inhibits cleavage of the phosphodiester bond on its 5'-side as well as on its 3'-side. Thus, because the fractionation especially of larger products is rather poor, and because even some smaller products run close together, one cannot use position as a means of identification and bands must always be analysed.

The blocking group is unstable at pH values above 10 so that the products may be analysed by alkaline hydrolysis (§ 4.4.1). However, this is less informative than removal of the blocking group with ammonia followed by subsequent analysis by digestion with P-RNase (§ 4.4.2.1) or U_2-RNase (§ 6.6) or other procedures. The conditions for removal of the blocking group with minimal phosphodiester bond cleavage are somewhat critical. The nucleotide is incubated in a sealed capillary tube with 10 μl of freshly prepared 0.2 N ammonia for 16 hr at room temperature. Alternatively, incubation is for about 4 hr at 37 °C. After the incubation the ammonia is allowed to evaporate from

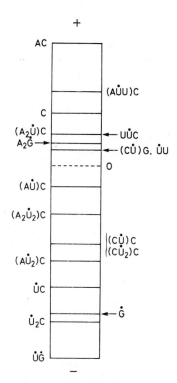

Fig. 6.3. Position of some common CMCT-blocked oligonucleotides in electrophoresis on Whatman paper at pH 3.5. The · signifies the presence of the carbodiimide reagent on the residue concerned. The position of spots varies somewhat so that bands must always be eluted to establish their sequence. (AU̇$_2$)C is often closer to U̇C than shown in the figure. The larger products are often rather streaky in comparison with the smaller products.

the sheet of polythene in a desiccator maintained under partial vacuum, and washed once with water and re-evaporated to ensure that no traces of ammonia remain. The dry nucleotide is now ready for analysis. A control may be run in any subsequent procedure to test for possible cleavage by the ammonia.

6.6 Cleavage next to A's with RNase U_2

This enzyme isolated by Arima et al. (1968) from the slime mould, *Ustilago*, is purine-specific if used in the correct proportion of enzyme to substrate. Thus, it may be used to analyse a T_1-end product, or fragments derived from T_1-end products where it cleaves specifically at *A residues*. Suitable conditions for oligonucleotides isolated from the standard two-dimensional system are as follows. Nucleotides are digested with approximately 10 µl of between 0.1 to 1.0 units/ml of enzyme (appendix 2) in 0.05 M sodium acetate, 0.002 M EDTA, *p*H 4.5 containing 0.1 mg/ml bovine serum albumin (appendix 2) for 2 hr at 37 °C. The digest may be fractionated by ionophoresis on DEAE-paper at *p*H 1.9, and the positions of some of the simpler products are shown in fig. 6.4. Notice that A, and products containing AA, such as

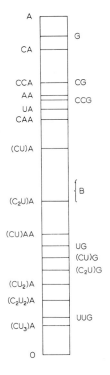

Fig. 6.4. Fractionation of products of RNase U_2 digestion of T_1-oligonucleotides by ionophoresis on DEAE-paper at *p*H 1.9. O is the origin and B the blue marker. Sometimes the faster A-ending products, e.g. CA and CCA move faster relative to G than is shown and may be ahead of G. These variations are probably caused by slight differences in the composition of the *p*H 1.9. mixture and in the temperature of the run.

CAA, are present, presumably because the A-A bond is partially resistant to cleavage by the enzyme. Spots isolated after 'homochromatography' on thin layers where a considerable amount of carrier oligonucleotide is present, need more enzyme. In this case, spots are digested with approximately 10 μl of 10 units enzyme/ml in the same buffer as above for 4 hr at 37 °C. Longer digestion times of up to 16 hr are required if the oligonucleotide is isolated by 'homochromatography' on DEAE-paper, as more carrier oligonucleotides are present than on the thin-layer system. The products of U_2-RNase digestion may be analysed for their composition by alkaline hydrolysis (§ 4.4.1.). If yields permit, or if necessary, other procedures such as 5'-end group determination (§ 4.4.2.4), partial P-RNase digestion (§ 6.8) or carbodiimide-blocking (§ 6.5) may be attempted.

6.7. Partial spleen acid RNase digestion
(Bernardi and Bernardi 1966)

The specificity of this enzyme is not accurately defined, but when used under partial digestion conditions it has a preference for purines over pyrimidines. When used on T_1-end products it therefore prefers A residues, although cleavage after C's and U's does occur. It is often possible to obtain a fairly specific cleavage in one or two positions only. Cleavage of A-A bonds is rare. The enzyme is rather inactive on a weight basis and high concentrations must be used. For oligonucleotides isolated from the standard two-dimensional system 10 μl of a 0.2 mg/ml solution of enzyme in 0.15 M sodium acetate, 0.01 M EDTA, pH 5.0 is used and digestion is for 2-3 hr at 37°C. Fractionation is best carried out on DEAE-paper using the 7% formic acid system as rather large products are to be separated from one another. For oligonucleotides isolated by 'homochromatography' on DEAE-paper 1 mg/ml of enzyme is used, and digestion is overnight at 37 °C. For nucleotides from thin layers, digestion with 1 mg/ml of enzyme for about 3 hr is suitable. Nevertheless, individual oligonucleotides will vary in their susceptibility and it is necessary to investigate a time course of digestion on portions of the sample, placing them side-by-side on the DEAE-paper. Sometimes the number of degradation products in an acid RNase digestion are rather too complex to be separated in one dimension and the standard two-dimensional ionophoretic fractionation system must be used. Products may be analysed for their composition (§ 4.4.1) and end group (§ 4.4.2.4). In addition, further enzymatic digestion using pancreatic or U_2-RNase may be used. The carbodiimide blocking procedure may also give useful results. It should always be considered that rarely will enough material be present for *all* of the above procedures and that the simplest should be attempted first and only if required should the other procedures be attempted.

6.8. Partial pancreatic RNase digestion

The conditions of digestion will vary for different oligonucleotides so that a trial experiment is necessary to establish suitable conditions. In order to control conditions accurately digestion is best carried out at 0 °C and in the presence of carrier tRNA (appendix 2) in the ratio of 100 parts by weight of carrier to 1 part by weight of enzyme. Thus an oligonucleotide isolated from the standard two-dimensional ionophoretic system may be digested for 30 min at 0 °C in approximately 10 μl of a solution of 0.1 mg of P-RNase/ml containing 10 mg of carrier tRNA/ml in 0.01 M Tris-chloride, 0.001 M EDTA, pH 7.5. The products may be separated by ionophoresis on DEAE-paper using 7% formic acid if the mixture of products in the digest is not too complicated. Alternatively, the digest may be fractionated on the standard two-dimensional system. In attempting partial P-RNase digests of oligonucleotides isolated by 'homochromatography' the weight of carrier oligonucleotide on the sample should be measured spectrophotometrically at 260 mμ (§ 2.2.2 footnote), and then that weight of P-RNase added in a volume of 10 μl of the above buffer at 0 °C which will give a weight ratio of 1 part of enzyme to 100 parts of substrate. Digestion is as before for 30 min at 0 °C. The weight of carrier usually isolated from an average sized spot fractionated by 'homochromatography' on the thin-layer system is about 150 μg.

Subject index p. 261

CHAPTER 7

Minor bases and tRNA

7.1. Introduction

The procedure and techniques described for studying the sequence of 5S RNA are immediately applicable to the sequence determination of tRNA. In fact tRNA, with a lower molecular weight than 5S RNA, is easier to sequence. Nevertheless the purification of a specific aminoacyl tRNA is a preliminary problem which may easily take longer to solve than the entire sequence analysis of the purified tRNA once it is obtained. Although it is not my intention to cover purification of tRNA, a subject fully described by Matthews and Gould (in prep.) in a companion volume of this series, a short description is given here (see § 7.2 below) summarising some of the special problems of purifying ^{32}P-labelled material.

Another difficulty in the sequence study of tRNA (and of rRNA) is the presence of minor bases. Many of the problems concerned with these compounds, including their isolation and identification, have already been discussed in general terms in § 2.6.1.9. However, in working with ^{32}P-labelled RNA there are special problems of identification because the usual method of establishing identity by a characteristic absorption spectrum cannot be used. Identity can only be established by a comparison of the mobility of the mononucleotide on paper electrophoresis and/or paper chromatography with authentic marker compounds. If markers are unavailable, e.g. if a new minor base is found, a radioactively labelled minor nucleotide is almost impossible to identify, although some of its properties such as methyl content and

pK of active groups may be deduced from its mobility relative to other compounds. However, once the fractionation systems are *calibrated* by marker compounds, the observed mobility values relative to an internal standard are as good an identification as a direct comparison with an authentic marker compound. These mobility values are thus an important means of identification and are discussed in detail in § 7.3 below. It is not intended to discuss the means of identification of all the minor nucleotides characterised in radioactive studies. However, a discussion of the nature of the evidence for those present in a tyrosine suppressor tRNA of *E. coli* (Goodman et al. 1970) will serve to illustrate some of the problems (see § 7.4).

Another problem in working with minor bases is that at least nine minor nucleotides are unstable to alkali (see table 2.3) so that it is necessary to use complete enzymic hydrolysis to isolate them in an unmodified form. The use of T_2-RNase for this purpose is, therefore, described (§ 7.5). One minor base, that is 4-thiouridine, gives more trouble than any other because at least in tRNA it is unstable to light. The nature of the reaction of s^4U in valine tRNA of *E. coli* was elucidated by Yaniv et al. (1969) using radioactive methods. It involves the formation of a cross-bridge of s^4U at nucleotide 8 with a C residue at nucleotide 13 from the 5′-end. As this property may be rather general in tRNA of *E. coli*, a discussion of the details of this and its significance is given at the end of this chapter.

7.2. *Purification of* ^{32}P-*tRNA*

The purification of ^{32}P-tRNA is not generally undertaken without some preliminary knowledge of how to purify the corresponding unlabelled tRNA. The reason for this is the difficulty of assaying specific tRNA molecules by their reaction with ^3H- or ^{14}C-labelled amino acids in the presence of very high amounts of ^{32}P. Many methods have been described for the purification of specific tRNA molecules but the introduction of the general method of Gillam et al. (1968) using benzoylated cellulose has considerably simplified the problem of the purification of any desired specific aminoacyl tRNA molecule. Once

Subject index p. 261

the procedure is established for the unlabelled tRNA it is applied to the ^{32}P-labelled material prepared as described in appendix 5. If very high specific activity ^{3}H-labelled amino acids (10 C/mmole) are available, it is just possible to follow the purification of a specific ^{32}P-labelled tRNA provided about 20 mg of carrier unlabelled RNA is added to the ^{32}P-labelled material. ^{3}H-counts are detected by liquid scintillation counting using machine settings which exclude ^{32}P from the ^{3}H-channel (see Dube et al. 1969). Once a fingerprint of a purified or partially purified tRNA is obtained it may be possible to purify it further by acrylamide gel electrophoresis using the vertical gel slab method described in Appendix § V.1.2.1. The various tRNA molecules are spread out over a distance of 5-10 cm (see appendix fig. A3) and for example a leucine tRNA in *E. coli* may be obtained almost pure from mixed tRNA by this method. Tyrosine tRNA of *E. coli* is also considerably purified by using this method.

7.3. *Minor bases*

At the time of writing the sequence of seven different tRNA molecules of *E. coli* have been established by radioactive methods (see table 7.1). All of them contain minor bases which are isolated in alkaline or T_2-RNase digests. The minor mononucleotides usually have slightly

TABLE 7.1
Known tRNA sequences of *E. coli*.

Sequence	Reference
Tyrosine and suppressor	Goodman et al. 1970
Methionine$_F$	Dube and Marcker 1969; Dube et al. 1969
Methionine$_M$	Cory and Marcker 1970
Valine$_1$	Yaniv and Barrell 1969
Phenylalanine	Barrell and Sanger 1969
Tryptophan and suppressor	Hirsh (1971)
Leucine and fragments	Dube et al. 1970

different mobilities from the four common mononucleotides on paper ionophoresis at pH 3.5. Methylated mononucleotides in general are slightly slower than the parent nucleotide. For example, ribothymidylic acid is slightly slower than uridylic acid. However, a number of minor compounds have similar mobilities either to one another (e.g. ribothymidylic acid and pseudouridylic acid) or to a parent nucleotide so that further characterisation by paper chromatography is necessary. After fractionation by paper ionophoresis at pH 3.5 bands are eluted with water into capillary tubes and dried down on polythene. Suitable

TABLE 7.2

Ionophoretic and chromatographic properties of some minor nucleoside 2′,3′-phosphates found in radioactive studies on tRNA and rRNA of *E. coli*.

Compound	Symbol	Electrophoretic mobility (pH 3.5) relative to Up†	Chromatographic mobility*	
			System 1	System 2
β-ureidopropionic acid	Y-	1.07	0.93 & 1.1	0.63
4-thiouridylic acid	s^4U-	~1.0		
2′-*O*-methyluridine 3′-phosphate	U^m-	1.0	1.12	1.3
5,6-dihydrouridylic acid	D-	1.0	0.9	
Ribo-thymidylic acid	T-	0.98	1.1	1.7
Pseudo-uridylic acid	Ψ-	0.98	0.79	0.63
Inosinic acid	I-	0.90	0.54	0.68
2′-*O*-methylguanylyl-guanylic acid	G^m-G-	0.89	0.22	0.20
4-amino-5(*N*-methyl)-formamidoisocytosine ribotide	'mG'	0.82	0.69	
2′-*O*-methylcytidylyluridylic acid	C^m-U-	0.79	0.67	~0.6
2′-*O*-methylguanosine 3′-phosphate	G^m-	0.78	0.82	0.90
N^2,N^2-dimethylguanylic acid	m_2^2G-	0.76	0.64	

(*continued*)

TABLE 7.2 (continued)

Compound	Symbol	Electrophoretic mobility (pH 3.5) relative to Up[†]	Chromatographic mobility[*] System 1	System 2
N^1-methylguanylic acid	m^1G-	0.74	0.70	
Guanosine 3'-phosphate	G-	0.74	0.50	0.66
N^2-methylguanylic acid	m^2G-	0.68	0.73	
N^6-dimethyladenylyl-(N^6-dimethyl) adenylic acid	m_2^6A-m_2^6A-	0.61		
2-methylthio-6-isopentenyl adenylic acid[**]	ms^2i^6A-	Just ahead of A-	1.2	Near front
Isopentenyl adenylic acid	i^6A-	Just ahead of A-	1.2	Near front
N^6-methyl adenylic acid	m^6A-	0.41	0.80	~1.6
Adenylic acid	A-	0.41	0.63	0.95
N^6-dimethyl adenylic acid	m_2^6A-	0.40	0.9	2.1
N^4-methyl-(2'-O-methyl) cytidylyl cytidylic acid	m^4C^m-C-	0.40 & 0.38	0.63	1.15
2'-O-methylcytidylyl cytidylic acid	C^m-C-	0.35	0.58	
Cytidylic acid	C-	0.21	0.74	0.98
2'-O-methyl cytosine-3'-phosphate	C^m-	0.20	0.94	1.15
5-methyl cytidylic acid	m^5C-	0.19	0.78	
N^7-methylguanylic acid	m^7G-	0.05	0.75	0.85 0.69
Unknown in tRNA^{Tyr} (E. coli)	G*-	0.05	0.57	0.83

[†] Electrophoretic mobilities are expressed relative to uridine 2',3'-phosphate = 1.00 on Whatman No.540 paper.

[*] Chromatographic mobility is for descending chromatography on Whatman No.1 paper. System 1 in propan-2-ol (680 ml), HCl (176 ml) and water to 1 litre (Wyatt 1951). System 2 in propan-2-ol (70 ml), water (30 ml), ammonia (1 ml) (Markham and Smith 1952).

[**] This compound is best distinguished from i^6A by electrophoresis on DEAE-paper at pH 3.5 (0.5% pyridine, 5% acetic acid) when it has an R_u of approximately 0.5 whereas i^6A has an R_u of approximately 0.9.

chromatographic systems are described in the footnote of table 7.2. In order to obtain reproducible and satisfactory results, samples must be loaded as narrow 2 cm bands on Whatman No. 1 paper at 10 cm from one end of the paper. 2-cm gaps are left between samples. The paper, loaded with the samples, is equilibrated in the tank for some hours with a solvent-saturated atmosphere (obtained by placing solvent in the bottom of the tank) before adding the solvent to the upper trough and starting the chromatography. It is normally complete, using *system 1*, in 24 hr, when the solvent front had almost reached the bottom of the paper. Nucleotides have relatively high R_f's in this system (Up has an R_f of approximately 0.8). In contrast, most nucleotides have very low R_f's in *system 2* and chromatography is carried out for 48 hr allowing solvent to drip off the bottom of the paper which is, therefore, serrated. However, when this is done, some very aliphatic A derivatives (e.g. ms^2i^6A and i^6A) which move near the solvent front, may be lost. These two compounds are in any case not separated by either of the chromatographic systems mentioned, and are best characterised by electrophoresis on DEAE-paper at pH 3.5 (see footnote to table 7.2). Mobilities are expressed relative to an internal standard, but even so variations do occur in the figures quoted, so that if possible markers should be run. These may either be radioactive markers, derived from minor nucleotides previously characterised or known, or authentic non-radioactive markers which are mixed with the unknown and detected after chromatography by ultraviolet absorption. The results are summarised in table 7.2 which is a composite of all the work on radioactive tRNA. It also includes minor nucleotides from 16S and 23S rRNA of *E. coli* (§ 8.4) and alkaline-resistant dinucleotides which have an *O*-methyl group on the 2'-position of ribose. The isomeric nucleoside 2'- and 3'- phosphates do not separate on the chromatographic systems, although a slight separation may be observed on paper electrophoresis. The mobility values of table 7.2 are the mid-point of the two isomers except where specified to the contrary. Thus when establishing the identity of an unknown minor nucleotide it is an advantage to characterise it as a mononucleoside 3'-phosphate, derived by T_2-RNase hydrolysis (§ 7.5, below).

Subject index p. 261

7.4. Identification of minor bases in tyrosine suppressor tRNA of E. coli

The sequence of the tyrosine suppressor tRNA of *E. coli* (Goodman et al. 1970) is shown in fig. 7.1 drawn in the accepted clover-leaf fashion.

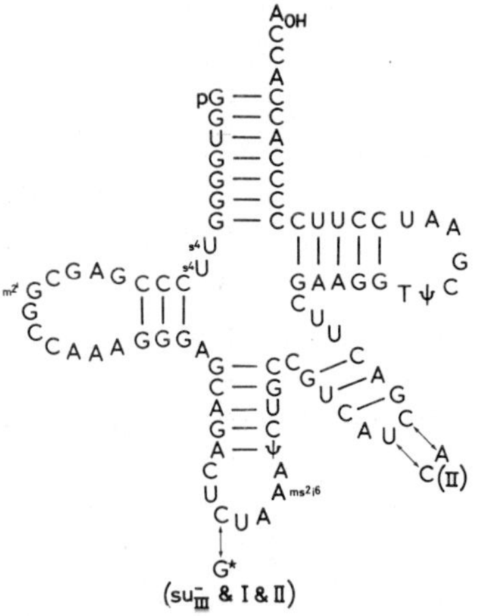

Fig. 7.1. Sequence of the su_{III}^+ (a tRNA^{Tyr}) of *E. coli*. C in the anticodon (bottom of fig) is replaced by a G* (an unknown derivative of G) in the su_{III}^- and wild type I and II tyrosine tRNAs. m$^{2'}$G indicates a 2'-O-methylated guanosine (position 17). The other symbols are standard – see table 7.2.

The main sequence shown in this figure is of the tRNA ^{Tyr} su_{III}^+ and contains two residues of both 4-thiouracil and pseudouracil, and one each of 2'-O-methylguanine, 2-methylthio-6-isopentenyladenine and ribothymine. The base change from C to G* in the anticodon, at the bottom of the figure, is found in both of the wild-type tyrosine tRNAs, called species I and III, as well as in a revertant of the suppressor, called su_{III}^-. The other two base changes in the 'finger' region of the tRNA of fig. 7.1 are found only in wild-type II tyrosine tRNA. Minor bases are identified with one exception as follows:

(1) *Tp* and *Ψp*. These are identified in the oligonucleotide TΨCG from the mobility values given in table 7.2. These values were derived from a study of the mobilities of marker non-radioactive Tp and Ψp

which were isolated from mixed tRNA and identified by their ultraviolet absorption spectra.

(2) *2'-O-methyl Gp*. This was characterised as the dinucleotide G^m-G- in alkaline digests of the oligonucleotide C-G^m-G-. This dinucleotide has very characteristic mobility values (table 7.2) and runs very slowly on chromatography. As no marker material was available, this dinucleotide was characterised as follows. With alkaline phosphatase it gave two radioactive products. One was inorganic phosphate and the other, in equal radioactive yield, gave pG on further digestion with venom phosphodiesterase and G^m- (table 7.2) on exhaustive digestion with spleen phosphodiesterase. It was, therefore, G^m-G.

(3) *2-methylthio 6-isopentenyl Ap*. This compound was discovered as a previously unknown mononucleotide which had electrophoretic properties similar but not identical to Ap and chromatographic properties, such that it had a faster R_f than any other known compound. It was, therefore, labelled A* and its position located in the pancreatic end product as the middle A (i.e. AA*AΨ) from the results of partial digestion with spleen phosphodiesterase. A* could be labelled in vivo with both $^{35}SO_4^{2-}$ and (methyl-^{3}H) methionine and the results suggested that A*p contained at least one atom of sulphur and one methyl group as secondary substituents. Nevertheless an identification was not possible at the radioactive level. However, when Burrows et al. (1968) described the presence of a 2-methylthio-6-isopentenyl A in *E. coli* tRNA, and Harada et al. (1968) located it in unlabelled tRNA^{Tyr}, these observations strongly suggested that A* was in fact this compound.

(4) *4-Thio Up*. This nucleotide was difficult to characterise because it was unstable to light. The only digestion product that could be found from the ^{32}P-labelled oligonucleotide believed to contain s^{4}U was U. Moreover the oligonucleotide was present in 0.1 molar yields with respect to other nucleotides and sequenced as UUCCCG. However, fingerprints of ^{35}S-labelled tRNA^{Tyr} (appendix § V.6) showed counts in the position of this oligonucleotide and no other in T_1-RNase digests, and in the position of GGGGU in the P-RNase digest. This was taken to indicate the presence of 4-thio U on the first of the two U

residues of the T_1-oligonucleotide. However, measurements of the total content of 4-thio U, which can be made spectrophotometrically (Lipsett 1965) on intact non-radioactive tyr$_I$ tRNA (Lipsett and Doctor 1967) strongly suggested that 2 moles of s^4U were present. As there was only one ^{35}S-labelled T_1-oligonucleotide one can only deduce that in fact both U's of the T_1-oligonucleotide may be modified in tyr tRNA (i.e. s^4Us^4UCCCG) but that this modification may be incomplete in the su_{III}^+ species under the conditions of its isolation as a ^{32}P-labelled molecule. The reason for the low yields of the presumed s^4Us^4UCCCG) may also be related to formation of cross-bridges as occurs in valine tRNA (§ 7.6).

(5) $G*p$. This nucleotide was isolated from su_{III}^- species (and species I and II) and had properties (table 7.2) differing from all other known compounds. Its electrophoretic properties suggested the presence of a basic group with a pK_a of between 5 and 6. From ribosomal binding studies on *E. coli* tyrosine tRNAs it was clear that G* (which occupied the first position of the anticodon – see fig. 7.1) had the same codon recognition properties as G. Small amounts of unlabelled G*p were then isolated from tRNA^{Tyr} and this compound was shown to have a similar, although not identical, ultraviolet absorption spectrum to Gp. This compound (which might be expected to have a basic group added to the imidazole ring of G at position 8) has still not been characterised and it would seem that this can only be done by such techniques as mass spectroscopy on the nucleoside isolated from unlabelled tRNA

In summary it is rather easy to characterise minor mononucleotides found in the radioactive studies by their mobilities on electrophoretic and chromatographic systems. This usually allows accurate comparison with known compounds and thus leads to an identification. However, if previously unknown compounds are isolated, these are not likely to be identified by radioactive work.

7.5. *Complete hydrolysis with T_2-RNase* (see also § 2.6.1.2.)

The characterisation of minor bases and especially of unstable ones, is probably the main reason for using this type of digest, rather than alkaline hydrolysis. Suitable conditions for the complete degradation of an oligonucleotide, or RNA to mononucleoside 3'-phosphate using a mixture of T_1-, pancreatic and T_2-RNase, are as follows. The substrate is digested for 2 hr at 37 °C with 10 μl of a solution of 2.0 units per ml of T_2-RNase (appendix 2) in 0.05 M ammonium acetate, pH 4.5 containing 0.05 mg of each of T_1- and P-RNase per ml. Fractionation may be carried out by ionophoresis on Whatman No. 540 paper at pH 3.5 as for an alkaline digest (§ 4.4.1). As only the 3'-phosphates are present as end products of digestion, rather than the mixture of the 2'- and 3'-phosphates as in an alkaline digest, better separations of closely related compounds may be observed. Thus, pseudouridine 3'-phosphate is more clearly separated from uridine 3'-phosphate than is the mixture of the 2'- and 3'-phosphates of pseudouridine separated from those of uridine.

7.6. A cross-linkage in E. coli $tRNA_1^{Val}$
(Yaniv et al. 1969)

A comparison of a T_1-RNase 'fingerprint' using the standard two-dimensional method of an irradiated versus a control ^{32}P-labelled $tRNA_1^{Val}$ showed the appearance of a new nucleotide,

$$\begin{array}{c} A\text{-}U\text{-}s^4U\text{-}A\text{-}G \\ | \\ C\text{-}U\text{-}C\text{-}A\text{-}G \end{array}$$

and a decreased yield of the two oligonucleotides making up the 'bridged compound' in the control, unirradiated sample. The sequence of this tRNA (Yaniv and Barrell 1969) clearly indicated that s^4U at position 8 and C at position 13 had been covalently linked although the chemistry of this linkage is not as yet known. Therefore, it is proposed that these two positions in the molecule are normally close to one another in the tertiary structure. This suggests that the 'dihydro U loop' of a transfer RNA lies close to and possibly parallel with the 'amino acid stem' of the tRNA. This feature is possibly general, as

Subject index p. 261

s⁴U is fairly common in *E. coli* tRNA molecules. Moreover, a similar linkage probably exists in tRNA^{Met_F} and tRNA^{Phe} of *E. coli* as a 'bridged' compound may be detected in low yields in unirradiated samples of these tRNAs.

CHAPTER 8

Limited sequence objectives: end groups of RNA and sequences adjacent to minor bases in ribosomal RNA

8.1. Introduction

In making a preliminary study of a high molecular weight RNA it is useful to establish the 5'-terminal and 3'-terminal sequences. Dahlberg (1968), for example, has established these sequences for RNA from some small bacteriophages using radioactive methods. A description of these methods, which are applicable to any uniformly ^{32}P-labelled RNA molecule, is therefore given below (§§ 8.2 and 8.3). Besides giving sequence information, studies such as these on end groups give information as to the homogeneity of the RNA. It should be emphasised that specific methods for detecting end groups are not usually required for studying molecules of the size of tRNA or 5S RNA. In these cases end groups are usually rather easily detected as end products of T_1-RNase digestion with rather characteristic properties (§ 5.2.3) on the fingerprint. The specific procedure is usually only required for studying high molecular weight RNA.

Another specialised application of the fingerprinting method (§ 4.3) is the isolation and sequence determination of oligonucleotides containing minor bases in rRNA of high molecular weight. Such molecules are too large to be immediately amenable to a total sequence determination so that the sequence of the regions containing minor bases serves as a useful preliminary probe. From such a study on 16S and 23S rRNA of *E. coli*, Fellner (1969) showed that there was signifi-

Subject index p. 261

cant 'clustering' of minor bases. It also emerged that all the major methylated oligonucleotide sequences in 23S RNA occurred *twice* in the molecule. A similar, but less extensive duplication of methylated sequences occurred in 16S RNA. The experimental approach used in this study, which again should be applicable to any molecule containing minor bases (but would be unnecessary in the case of a purified tRNA) is also described below (§ 8.4).

8.2. End groups in RNA

These may be isolated specifically because the *5'-end* of all known RNA molecules is characterised by the presence of at least one phosphate group, and the *3'-end* by the presence of a free hydroxyl group (e.g. the molecules have the form $(pp)pN(pN)_n pN_{OH}$) where N = any nucleoside and n is a variable but large number of bases). The form results from the mechanism of synthesis of RNA which is by condensation of nucleoside 5'-triphosphates. A few nucleic acids, such as the small RNA viruses, retain their 5'-terminal triphosphates. However, in most cases the two phosphoanhydride linkages are removed by phosphatase leaving only a 5'-terminal *mono*-phosphate ester. It is not my intention to review all the methods that have been devised for determination of end groups. I wish merely to illustrate those which have been successfully applied in studies on radioactive RNA.

8.2.1. 5'-end nucleotide

The 5'-terminal nucleotide of a radioactive RNA is released by alkaline hydrolysis in the form (pp)pNp whereas all other internal nucleotides are liberated as Np. This terminal grouping may be separated from the four mononucleotides and characterised by running on a two-dimensional system similar to that used by Dahlberg (1968) as follows. The uniformly ^{32}P-labelled RNA is hydrolysed with alkali (§ 4.4.1) and fractionated by paper ionophoresis on Whatman No. 540 paper at pH 3.5 using a buffer (§ 3.6) to which 0.001 M EDTA has been added. Without EDTA nucleoside 3',5'-diphosphates streak badly, presumably because of binding to metal ions in the paper. As compounds such as pppNp have a mobility approximately 1.6 times that of Up on this system care must be taken not to run them off the end of the paper. The products are then fractionated by ionophoresis on DEAE-paper using 7% formic acid or the pH 1.9 system in a *second dimension*. Nucleotides are transferred to the DEAE-paper by sewing a 3 cm wide strip of the No. 540 paper onto the DEAE-paper, 10 cm from one end.

End groups with their higher numbers of phosphate residues are more negatively charged and run correspondingly faster than mononucleotides in the first dimension but slower in the second dimension

TABLE 8.1
Mobilities of nucleoside 3′, 5′-diphosphates.

Material	R_u on ionophoresis on No. 540 paper at pH 3.5	R_G^* on ionophoresis on DEAE-paper pH 1.9
pCp	1.0	0.9[†]
pAp	1.05	0.81**
pGp	1.3	0.34**
pUp	1.6	0.42

* R_G = mobility relative to Gp = 1.00.
** Average of position of the 2′- and 3′-phosphate isomers which show slight separation.
[†] The R_G may be somewhat higher and similar to Gp under some conditions so that it may be necessary to use DEAE-ionophoresis at pH 3.5 to separate pCp from Gp.

of the system described above. The mobilities of the 4 nucleoside 3′,5′-diphosphates are shown in table 8.1. Authentic radioactive markers may be synthesised from unlabelled dinucleoside monophosphates (i.e. NpN) by reaction with ^{32}P-γ-labelled ATP and polynucleotide phosphokinase (§ 9.2), followed by alkaline hydrolysis. Alternatively a mixture of the four 3′, 5′-diphosphates may be simply obtained by partial digestion of any uniformly ^{32}P-labelled RNA with snake venom phosphodiesterase or with *Neurospora crassa* endonuclease (Linn and Lehman 1965) under conditions which cleave internally to give 5′-phosphate bearing oligonucleotides. The oligonucleotides are then subjected to alkaline hydrolysis without intermediate fractionation to give the nucleoside diphosphates in good yield in addition to the mononucleotides. Products are conveniently fractionated by ionophoresis on DEAE-paper at pH 1.9. Independent evidence for the correctness of the identification shown in table 8.1 is also derived by the study of tRNA which gives predominantly pGp, 5S RNA of *E. coli* giving pUp (§ 5.2.3) and 6S RNA of *E. coli* giving predominantly pAp (Brownlee 1971).

The *approximate* positions on DEAE-cellulose ionophoresis at pH 1.9 of compounds of the form *ppNp* may be found by comparison with authentic unlabelled compounds of the form pppN, which have a similar number of phosphate groups. However, such end groups, which do not commonly appear to be present in RNA, should also be characterised by degradation to the diphosphate pNp and inorganic phosphate. This may be achieved by using the enzyme potato apyrase as described by De Wachter et al. (1968a).

Compounds of the form *pppNp*, such as the pppGp found in f_2 bacteriophage RNA, may be characterised by a specific degradation to pp | pNp using high concentrations of snake venom phosphodiesterase. Inorganic pyrophosphate runs faster than inorganic phosphate on ionophoresis on paper at pH 3.5 (Roblin 1968). The yield of an end group may easily be determined by counting the spots isolated on the above two-dimensional system and comparing with the yields of the mononucleotides. Sometimes partial products of alkaline hydrolysis are present and, for example, products like UpUp might be confused with pUp. However, the former may be distinguished as it degrades on further alkaline hydrolysis and, moreover, gives UpU + p with alkaline phosphatase. pUp (and any other 5'-end product) gives p as the only radioactive product on degradation with alkaline phosphatase.

8.2.2. *5'-end oligonucleotide*

With a knowledge of the 5'-terminal mononucleotide released by alkaline hydrolysis it is possible to use this as an assay in order to screen all oligonucleotides present in a fingerprint of an enzymic digest of the RNA. Thus if pGp is deduced from alkaline hydrolysis it is only necessary to isolate this product from an oligonucleotide present in a complete P-RNase digest fractionated on the standard two-dimensional system. Nevertheless, specific procedures for isolation of 5′-terminal oligonucleotides have been devised and De Wachter et al. (1968b) with the knowledge that the 5′ mononucleotide of MS2 RNA (strain similar to f_2) was pppGp, prepared a P-RNase digest of ^{32}P-RNA and fractionated the products by DEAE-cellulose column chromatography in 7 M urea at pH 7.9 (§ 2.3.1). Under these conditions pppGp moves with the tetranucleotides and thus an oligonucleotide containing it would be expected to run in the penta- or higher nucleotide peaks. Alkaline hydrolysis was used to screen the peaks for the presence of pppGp, which was found in the heptanucleotide fraction. This suggested a sequence containing three more negative charges than pppGp, i.e. pppGpPupPupPyp. This product was separated from the other components in the heptanucleotide fraction by a gradient chromatographic system on DEAE-paper at pH 8.0 in 8 M formamide and analysed by alkaline hydrolysis to give the sequence pppGpGpGpUp. Other fractionation methods such as the two-dimensional method described in §§ 5.5.1 or 5.5.2 might have been used to isolate such a sequence from the heptanucleotide fraction. Alternatively the heptanucleotide fraction may be treated with alkaline phosphatase and rerun on the DEAE-7 M urea fractionation system. The end sequence should now be found as the only product in the *trinucleotide region* as it should lose four phosphate residues, whereas all other heptanucleotides should lose only a single phosphate residue and now chromatograph with the *hexanucleotides*. Although this latter approach offers a specific method of isolation it has the disadvantage of destroying the characteristic pppG end-group. Thus, if there are any endonuclease contaminants present during the phosphatase digestion artefacts could appear in the trinucleotide region leading to an incorrect identification.

Subject index p. 261

8.3. 3'-terminal oligonucleotide – a 'diagonal' method
(Dahlberg 1968)

This specific method depends on the fact that in a complete T_1-RNase digest of an RNA the only product which is not susceptible to degradation with alkaline phosphatase is the oligonucleotide derived from the 3'-terminal end. All other oligonucleotides have a 3'-terminal phosphate and are susceptible. Thus, if an enzyme digest of an RNA is fractionated in one dimension followed by treatment with alkaline phosphatase before running at right angles to the first dimension but using the same fractionation method, the only product remaining on the 'diagonal' should be the 3'-terminal oligonucleotide. To illustrate this the procedure of Dahlberg (1968), who devised the method for studying bacteriophage RNA's, is described in some detail. A T_1-RNase digest of the ^{32}P-labelled bacteriophage f_2-RNA is applied as a spot to a strip of DEAE-paper and fractionated by ionophoresis using 7% formic acid. The nucleotides are located by radioautography and a strip (usually 2 cm wide) containing them is cut out. A solution of freshly dissolved alkaline phosphatase (0.6 mg/ml in 0.5 M (NH_4)-HCO_3, 0.025 M $MgSO_4$, 0.0001 M $ZnSO_4$) is then applied to the area of the paper containing the oligonucleotides with a disposable Pasteur pipette. (About 1 ml of solution was usually required.) The paper strip is placed on its side on a glass rack and incubated at 37°C for 2 hr in a water-saturated atmosphere inside a desiccator. After 2 hr salt was then washed off with water and the strip of paper allowed to dry. It was then sewed onto a sheet of DEAE-paper (85 × 47 cm) at 10 cm from one of the short ends of the paper, and after cutting out the extra paper behind the new origin, ionophoresis was carried out at right angles to the first dimension using 7% formic acid, as before. A radioautograph using this method applied to bacteriophage f_2-RNA is shown in fig. 8.1. The origin is at top left, the first dimension being from left to right, the second being from top to bottom. The position of the blue marker (appendix 2) run on the first dimension is used to mark the diagonal indicated by the dashed line on the Plate. Although there are a number of radioactive spots lying in this diagonal, only the

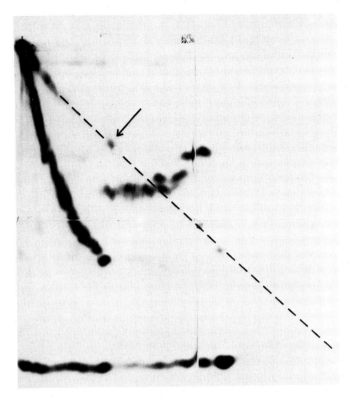

Fig. 8.1. A diagonal radioautograph (DEAE-paper ionophoresis using 7% formic acid) of a T_1-RNase, treated with alkaline phosphatase between the 2 runs. For further details see text. (From Dahlberg 1968.)

spot marked with an arrow lacked a Gp on further analysis. Its sequence was established by standard methods and was established as UUACCACCCA$_{OH}$. The other products on the diagonal are probably nucleotides terminating in a 2′, 3′-cyclic phosphate which makes them resistant to alkaline phosphatase treatment. In addition a line of dephosphorylated oligonucleotides crosses the diagonal. In order that artefact end groups are not produced it is important that the T_1-RNase is free of any contaminating phosphatase. This is especially true as the required conditions of digestion are reasonably vigorous, such that cyclic phosphates are absent (or in very low yield). A solution of T_1-RNase (1 mg/ml in 0.01 M Tris-chloride, 0.02 M EDTA, pH 7.2) is acidified by the addition of 0.05 volumes of 1 N HCl and incubated for 10 min at 22 °C; the solution is then neutralised by the addition of 0.1 volume of 1 M Tris-chloride, pH 7.2 plus 0.05 volume of 1 N NaOH and should be free from phosphatase.

Once the mobility of the 3'-terminal oligonucleotide is known on the fractionation system (DEAE-cellulose ionophoresis, using 7% formic acid) it may in fact be more convenient to purify it by using the standard two-dimensional fractionation system. The mobility of such nucleotides is usually rather slower than other products in the first dimension because of the absence of a 3'-terminal phosphate group.

8.4. Sequence around methylated bases in 16S and 23S RNA of E. coli

The complexity of a fingerprint of uniformly ^{32}P-labelled high molecular weight rRNA is such (see fig. 4.1) that it would require an exhaustive analysis of all the oligonucleotides in order to detect those containing minor bases. Moreover, in such an approach there is considerable chance that many such oligonucleotides would be overlooked. In order to locate minor bases among the many oligonucleotides present in a fingerprint, rRNA is prepared labelled specifically with ^{14}C in the methyl groups of minor bases using (^{14}C-methyl) methionine as a methyl donor as described in appendix §V.5. This marks all the known minor bases in rRNA with the exception of pseudouracil. After separation of 16S and 23S rRNA by density-gradient centrifugation (appendix § V.1.1) a fingerprint of (^{14}C-methyl) rRNA is prepared by the combined action of T_1-RNase and alkaline phosphatase. Another fingerprint of ^{32}P-labelled rRNA (appendix § V.1.1) is also prepared and run under identical conditions. Fellner and Sanger (1968) found that the radioautograph of the fingerprint of 23S ^{14}C-methyl rRNA, which required up to one week's exposure, gave 12 spots in good yield with a few others in lower yield. The minor nucleotides were liberated from these labelled oligonucleotides by alkaline hydrolysis or complete hydrolysis with venom phosphodiesterase and tentatively identified from their electrophoretic and chromatographic mobilities (see § 7.3). Confirmation of the identification was obtained in most cases by co-chromatography and co-electrophoresis with authentic non-radioactive marker compounds. As these were usually only available as the free base (in the case of purines) or a nucleoside (in the case of pyrimidines) it was necessary to treat suspected ^{14}C-labelled purine nucleotides with acid to hydrolyse the glycosidic bond to give the free base. Suspected pyrimidine nucleotides were treated with an excess of alkaline phosphatase to give nucleosides.

On the basis of the position of the ^{14}C-labelled oligonucleotides on the fingerprint it is possible to treat the corresponding oligonucleotide on the ^{32}P-labelled fingerprint run under identical conditions. In the case of ambiguity a study of the products of alkaline hydrolysis will usually confirm whether the correct minor nucleotide is present. A more accurate comparison of fingerprints, however, can be made as follows. ^{32}P- and ^{14}C-methyl-labelled 23S rRNA are mixed in a count ratio of approximately 100 to 1 and a fingerprint prepared and exposed. The radioautograph obtained will then show the position of the ^{32}P-labelled oligonucleotides which will mask the much lower amounts of ^{14}C-methyl-labelled oligonucleotides. The fingerprint is then stored for about six half lives (3 months) when the ratio of ^{32}P- to ^{14}C-counts will have decreased to about 1.5 to 1. Re-exposure of the fingerprint will now give a radioautograph on which the ^{14}C-methyl-labelled products are clearly visible. Superimposition of the two radioautographs should thus allow an accurate location of ^{32}P-oligonucleotides which contain minor bases. Feller (1969) described another method of locating oligonucleotides containing minor bases on the fingerprint, which has the advantage of not requiring a three months' wait. (^{3}H-methyl) rRNA is prepared (appendix § V.5) and mixed with ^{32}P-rRNA. Methylated spots are detected on the fingerprint of the doubly-labelled RNA by counting circles of DEAE-paper corresponding to the ^{32}P-labelled oligonucleotides specifically for ^{3}H using a liquid scintillation counter (§ 5.4).

Subject index p. 261

The sequence of these oligonucleotides was carried out on the ^{32}P-labelled oligonucleotides. Mostly they were present as pure or nearly pure spots on the fingerprint. Spot 11 of table 8.2 was an exception to this and was severely cross-contaminated with other products. The results of this sequence analysis, which made extensive use of the method of partial digestion with snake venom phosphodiesterase, on the products from 23S rRNA of *E. coli* are shown in table 8.2. The molar

TABLE 8.2

Sequences and yields of major methylated oligonucleotides from 23S RNA of *E. coli* (from Fellner and Sanger 1968).*

Sequence (end products of T_1-RNase and alkaline phosphatase)	Relative molar ratios	Suggested frequency moles/mole of RNA
1. m^{2}G	1.2	2
2. C-Um-G	0.9	2
3. C-(C,U)m^{2}G	1.1	2
4. C-m^{2}A-U-G	1.05	2
5. A-U-m^{5}C-C-G	1.2	2
6. C-m^{6}A-A-G	1.2	2
7. A-C-C-m^{6}A-G	0.9	2
8. Ψ-A-A-C-mU-A-Ψ-C-G†	0.9	2
9. A-C-A-U-A-U-m^{1}G-Ψ-T-G	1.0	2
10. A-A-A-T-U-C-C-U-U-G	1.0	2
11. - - -m^{7}G- - - - - - G	0.85	2

* Some minor products present in low yield and quoted by Fellner and Sanger (1968) have been omitted from this table for clarity.
† mU is an unknown derivative of U.

ratios are derived by measuring the yield of ^{14}C-methyl-labelled oligonucleotides (§ 5.4). The observed number of counts is divided by the number of methyl groups in the oligonucleotide and the resultant figure normalised with respect to nucleotide no. 11 which is arbitrarily given a value of 1.00.

In order to obtain the *absolute* frequency of occurrence of these sequences in rRNA it is necessary to determine the total number of methyl groups per molecule of 23S rRNA. An accurate method of doing this is to compare the ratio of counts (i.e. ^{32}P to ^{14}C ratio) of a mixture of ^{32}P- and ^{14}C-methyl rRNA with the same ratio determined for a single methyl-labelled mononucleotide isolated from a purified oligonucleotide. The procedure is as follows. ^{32}P-labelled and ^{14}C-methyl-labelled rRNA are prepared separately and mixed to give a ratio of ^{32}P to ^{14}C in the rRNA of approximately 10 to 1. A fingerprint of this material is prepared and two large oligonucleotides characteristic of 23S RNA (nucleotides 9 and 11 of table 8.2) are eluted and hydrolysed with alkali. The mononucleotides m^{1}G and 'mG' (the alkaline degradation product of m^{7}G) derived from nucleotides 9 and 11 respectively were purified by electrophoresis and the ratio of

TABLE 8.3
Determination of numbers of methyl groups in 23S rRNA of *E. coli* by double-label counting (Fellner and Sanger 1968).

Material	^{32}P/^{14}C*	Methyl groups/molecule ($x/y \times N$**)
23S RNA	7.7 (y)	—
m^{1}G from nucleotide 9 of table 8.2	0.06 (x)	28
mG† from nucleotide 11 of table 8.2	0.06 (x)	28

* Average of two experiments.
** N is the number of nucleotides per 23S RNA = 3590.
† See table 7.2 for identity of mG, the product of alkaline degradation of m^{7}G.

^{32}P to ^{14}C counts in them measured by counting in a liquid scintillation counter. The results of this experiment are given in table 8.3. From these results the total numbers of methyl groups is calculated as follows. If 0.06 (the observed ratio) is the specific activity corresponding to a single methyl group for single phosphate-labelled mononucleotides there would have to be an average of 0.06 methyl groups per 7.7 (i.e. 1 per 130) non-methylated residues to account for the observed difference in specific activity in the whole molecule. As the chain length of 23S RNA (deduced from the composition and the molecular weight, is 3590, the total number of methyl groups is 3590/130, which equals 28. On the basis of this result the relative molar ratios must be doubled. The nucleotides in table 8.2. will then account for 24 out of the total of 28 methyl groups. The remainder are probably accounted for by the oligonucleotides present in lower yields which were not included in table 8.2.

CHAPTER 9

In vitro labelled ^{32}P-RNA

9.1. Introduction

In previous chapters I have described methodology for sequencing uniformly ^{32}P-labelled RNA. The use of paper fractionation systems, in particular two-dimensional systems, combined with the sensitivity of autoradiography for the detection of ^{32}P, together provide a high-resolution and yet simple method for sequencing RNA. Nevertheless, it is not always possible to obtain highly radioactive RNA. For example mammalian RNA from a particular tissue cannot be labelled to a very high specific activity because of the limited amount of ^{32}P-phosphate which may be injected into an animal. In addition and of more significance, a high 'pool' of phosphate in the body fluids effectively dilutes the specific activity of the ^{32}P-phosphate that is injected. This difficulty may be overcome if the particular cell line will grow in tissue culture, or at least in a suspension culture, in a low phosphate-containing medium when it may be labelled in vitro (see appendix § V.3). However, few cells other than embryonic or malignant cells grow well under these conditions. It is thus of some general significance to establish methods for sequencing unlabelled RNA.

One method, which is of general application, is to prepare a RNase digest of an unlabelled RNA preparation and then subsequently label the oligonucleotides which are formed at their free 5′-hydroxyl group with a ^{32}P-phosphate ester group. This may be achieved by using an enzyme from the T4-bacteriophage – polynucleotide phosphokinase

(Richardson 1965) and γ-^{32}P-labelled ATP as the ^{32}P-phosphate donor. The resultant 5'-terminally ^{32}P-labelled oligonucleotides may then be fractionated by the standard methods developed for studying *uniformly* ^{32}P-labelled oligonucleotides. The determination of the sequence of such labelled fragments is, however, limited by the fact that only the 5'-terminal nucleotide is labelled. This procedure was developed by Szekely and Sanger (1969) from a labelling procedure of Takanami (1967) which he had used for the 5'-end group determination of unlabelled intact RNA. Further details of this method, including its applications and limitations, appear below (§ 9.2).

A second method of studying the sequence of unlabelled RNA is to synthesize a ^{32}P-labelled copy of it by using the unlabelled material as a template for a specific replicase, in the presence of α-^{32}P-phosphate labelled nucleotide triphosphates as substrates. Because of the known accuracy of the enzymatic copying process, the product RNA has an exactly complimentary sequence to that of the template, and is thus a valid method of sequencing the original unlabelled material. Moreover, because enzymatic synthesis of RNA proceeds from a single starting point on the template, in a specified direction, it is possible (under conditions of synchronised synthesis) to specifically label sections of the molecule at the 5'-end (the end of RNA synthesized first). Such an approach has been used with notable success by Billeter et al. (1969) for studying the first 175 nucleotides from the 5'-end of the RNA of the bacteriophage Qβ. Some details of this approach are discussed below in § 9.3. Although this method is of less general application than the phosphokinase method (§ 9.2) it may become more useful as our knowledge of controlled enzymatic copying of RNA (and DNA) becomes more precise. For example, in order to apply these methods, it is necessary that correct initiation of RNA synthesis occurs. It is perhaps not entirely out of place in this monograph to speculate that in fact the sequence of a specific piece of even DNA may in the future be achieved by sequencing a RNA copy of it made in vitro.

Other methods of radioactive labelling of RNA appear in the literature, for example, the use of ^{3}H-borohydride to label periodate-treated oligonucleotides (e.g. De Wachter and Fiers 1967). Nevertheless ^{3}H is not (or very poorly) detected by radioautography and so is unsuitable for the fractionation methods developed for ^{32}P-labelled material.

9.2. Polynucleotide phosphokinase labelling

Szekely and Sanger (1969) and Labrie and Sanger (1969) have described how to prepare fingerprints of unlabelled tRNA and haemoglobin mRNA, respectively, using this method. Very high specific activity γ-^{32}P-ATP is required for the successful application of the method and this may be prepared according to Glynn and Chappell (1964) from carrier-free ^{32}P-phosphate (see recipe on p. 608 of Szekely and Sanger 1969), or alternatively purchased from the Radiochemical Centre, Amersham. Commercial material should be purified by electrophoresis on DEAE-paper using 7% formic acid (§ 4.3). The specific activity should be at least 10 mC/μM. 5'-hydroxylpolynucleotide kinase is prepared according to Richardson (1965). It may be stored at 0 °C in the solution eluting from the final chromatographic separation or in 50% glycerol at -20 °C for a few months. Further details appear in Szekely and Sanger (1969).

The labelling reaction is carried out as follows (Labrie, unpublished). 1 μg is digested with T_1-RNase and bacterial alkaline phosphatase together (§ 4.2) in a small volume – usually 20 μl. After digestion the enzymes are extracted into phenol by mixing with an equal volume of water-saturated phenol and separating the layers by centrifugation in a capillary tube. The aqueous phase is then recovered and freeze-dried to remove traces of phenol. To the dried-down aqueous phase is then added about 5 μl (the exact amount will depend on the activity of the enzyme) of the phosphokinase, 1 μl of buffer (0.05 M $MgCl_2$, 0.05 M mercaptoethanol, and 0.1 M Tris-chloride, pH 7.4) and 1 μl of ^{32}P-ATP containing about 2 mμM (about 20 μC). This is a two- to three-fold molar excess of ATP to oligonucleotides. The mixture is incubated at 37 °C for 30 min. It is usually advantageous for the subsequent fractionation procedure to degrade excess ^{32}P-ATP into ^{32}P-phosphate with an ATPase. This is because the ^{32}P-ATP streaks in the first dimension of the standard two-dimensional system and may

228   SEQUENCES IN RNA

obscure some of the labelled oligonucleotide. ^{32}P-phosphate, however, is less serious as it occupies a position on the fingerprint well away from any oligonucleotides. Myosin may be used and 2 μl of a suspension of 2 mg/ml in 0.1 M EDTA (neutralised) is a suitable amount to add to the incubation mixture containing the ^{32}P-ATP and incubation continued for a further 10 min at 37°C. The material is fractionated by using the standard two-dimensional ionophoretic system.

An example of a fingerprint of *rabbit* reticulocyte 5S RNA (unpublished work of Dr. Labrie) prepared by this method is shown in fig. 9.1 on the left. On the right of the plate is shown the system of numbering of the products. The sequence of these products may be deduced

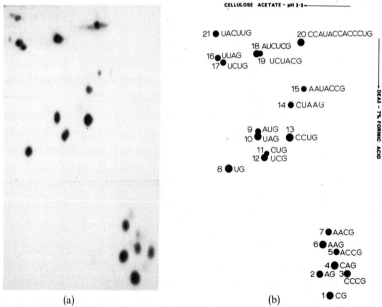

Fig. 9.1. Two-dimensional fingerprint of a combined T$_1$-RNase and alkaline phosphatase digest of rabbit reticulocyte 5S RNA radioactively labelled with a 5'-phosphate using polynucleotide phosphokinase. (a) is a radioautograph and (b) is a line diagram showing the sequence of spots. The sequences, but not the yields, are identical to those of Forget and Weissman (1968) for human 5S RNA. (From Labrie, unpublished results.)

from the known relationship of position to composition (§ 4.5.1) and from a knowledge of the sequences previously found in human 5S RNA by Forget and Weissman (1968) using uniformly ^{32}P-labelled methods. In fact a comparison of fingerprints shows no differences in the 5S RNA of humans and rabbits. These sequences are given on the right-hand side of fig. 9.1. Further examination of the fingerprint shows there is no unchanged ^{32}P-ATP on the fingerprint except for a contaminant in the ATP preparation which streaks in the first dimension through spot 20. All the ATP has been converted to ^{32}P-phosphate (which has run off the fingerprint) by the ATPase activity of the myosin. The determination of the sequence of the oligonucleotides, however, presents difficulties as most sequence procedures developed for uniformly ^{32}P-labelled oligonucleotides are not applicable to oligonucleotide labelled solely in the 5'-terminal position. Nevertheless the 5'-terminal mononucleotide may be easily found as the only radioactive product in either an alkaline hydrolysate (§ 4.4.1) as a mononucleoside 3', 5'-diphosphate or in a complete digestion with venom phosphodiesterase (§ 4.4.2.4) as a mononucleoside 5'-phosphate. In order to develop methods for finding the internal sequence Labrie (unpublished work) used the oligonucleotides of *known* sequence in fig. 9.1 and applied the method of partial digestion with venom phosphodiesterase (§ 5.2.2) which degrades oligonucleotides from the 3'- end leaving a series of 5'-phosphate-labelled intermediates. Examples of such digests fractionated by ionophoresis on DEAE-paper at *p*H 1.9 are shown in fig. 9.2. Thus the position of some simple di- and trinucleotide sequences in this system was established. For example, in spot 5 of fig. 9.2 (-A-C-C-G) unchanged material runs just ahead of the blue (B) marker. The first product running about twice as fast is -A-C-C. The next faint product running just ahead of -A-C-C is -A-C whilst the final (strong) product is in the position of pA (a very small amount of a contaminant pC is also present ahead of pA).

In applying this method to *unknown* sequences, a knowledge of a probable composition (as well as 5'-end group) of the oligonucleotide from its position on the fingerprint is of considerable help in interpreting the results of the partial digestion with venom phosphodiesterase.

Subject index p. 261

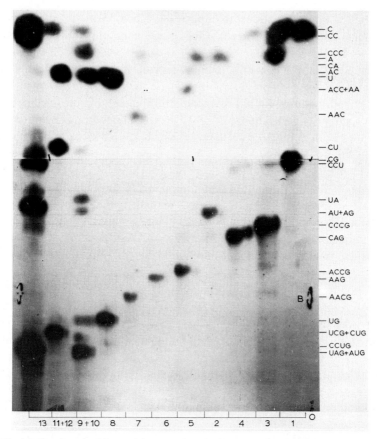

Fig. 9.2. Fractionation of a partial venom phosphodiesterase digest of some 5'-phosphate labelled oligonucleotides from rabbit reticulocyte 5S RNA by ionophoresis on DEAE-paper at pH 1.9. A control and several different digestion conditions of the sample were applied at O, the numbers being as in fig. 9.1. The various degradations were interpreted as shown by the sequences on the right of the figure. Note they all have a 5'-phosphate and a 3'-hydroxyl group. B is the blue marker. (From Labrie, unpublished results.)

The M values of the degradations may also be useful especially for the *larger* oligonucleotides and the values for C, A and U (but not G) quoted in table 4.1 apply. However, exceptions occur for smaller nucleotides (see below). As an example the M values measured for spot 5 on fig. 9.2 are 1.2, 0.06, 0.04, respectively. These correspond to the successive removal of a G, C and C which agrees with the sequence ACCG. In fact, the M value for G is rather low, but there is no doubt of this as G must always be 3′-terminal. The reader is invited to make a similar analysis of the other degradations in fig. 9.2. It should be emphasized that contaminants or mixtures may be present which can confuse the interpretation. Thus the first two products of the degradation (9 + 10) cannot be the correct degradation products as the loss of a pG must at least *double* the mobility of the product. The presence of two further products close together suggests a mixture which is confirmed by the presence of both pA and pU as the major end products of digestion. The contaminant pC is also present in lower amounts. Both the position of the original oligonucleotide on the 'fingerprint' and the degradation support the result of 9 + 10 being -A-U-G and -U-A-G, although the reader will notice that the M value for the cleavage of A and U are both about 0.5 and thus disagree with the values in table 4.1. This illustrates that *small* oligonucleotides in particular contradict these M values. It is obvious that with a single sequence procedure some care is needed in interpretation of the results and it is probable that other methods for sequencing of these oligonucleotides need to be developed. The fingerprint of fig. 9.1 shows rather low yields of some oligonucleotides, e.g. spots 5 and 7. This could well be due to incomplete phosphorylation which is a rather serious limitation. In summary this method is probably developed sufficiently to be a useful 'fingerprinting' method for comparison of RNA molecules. It remains to be demonstrated that it is possible to deduce a complete sequence by using this method.

9.3. Synchronised synthesis of RNA using purified replicase

For a number of technical reasons, which need not be discussed here, it is possible to obtain in vitro a synchronised synthesis of a segment of $Q\beta$ bacteriophage RNA from the 5′-end of the molecule and extending about 175 residues into the chain (Billeter et al. 1969). The reaction may be carried out with all four substrates (the four nucleoside triphosphates) labelled with ^{32}P in the α-position (all at the same specific activity). The synthesized material is then susceptible to 'fingerprinting' and analysis by the usual methods for uniformly labelled RNA. However, with this in vitro system it is also possible to use each of the four labelled triphosphate substrates *separately* to obtain 'nearest-neighbour' sequence information, which is extremely valuable and consider-

TABLE 9.1
Derivation of the sequence (G)UUUCCAG[A].[†]

Experiment	Labelled substrate	Information		Deduction
		Product of		
		Alkaline hydrolysis	Pancreatic RNase	
1.	ppp*U ppp*U ppp*C ppp*G	2.7 Up*, 1.0 Ap* 1.8 Cp*, 1.0 Gp*	Ap*Gp*, Up*, Cp*	(U$_3$C$_2$)p*<u>Ap*Gp*</u>
2.	ppp*A	1.2 Cp*, 1.0 Gp*	ApGp*, Cp*	(U$_3$C)p<u>Cp*ApGp*</u>[A]
3.	ppp*U	Up*	Up*	<u>Up*</u>Up*UpCpCpApGp [A]
4.	ppp*C	1.0 Up*, 1.0 Cp*	Up*, Cp*	UpUp<u>Up*Cp*Cp</u>ApGp [A]
5.	ppp*G	Ap*	Ap*Gp	UpUpUpCpCp<u>Ap*Gp</u> [A]

[†] The asterisk marks the labelled ^{32}P-phosphate. An underlined part of a sequence is that which can be deduced from the information in that single experiment (horizontal row). The square brackets are the 3' nearest neighbours of the G. Figures are quantitative values estimated by liquid scintillation counting.

ably simplifies the problem of sequencing the T_1-end products. The derivation of the sequence of one T_1-end product UUUCCAG will be considered in detail to show how this 'nearest-neighbour' information is used to deduce the sequence. The results of five separate labelling conditions are recorded in table 9.1. They differ in the input label which in experiment 1 is all four nucleoside triphosphates (at the same specific activity). However, in experiments 2-5 each of the substrates is used in turn, in the presence of the other three *unlabelled* substrates at the same concentration. In each experiment the labelling reaction is for 30 sec using the template Qβ RNA and the purified polymerase. The radioactive products synthesised in this time period are purified, digested with T_1-RNase and five separate fingerprints prepared using the standard two-dimensional fractionation system. Table 9.1 gives the results of the analysis of one of the spots which is in the same position in all five fingerprints. Experiment 1 gives the results of alkaline hydrolysis and pancreatic RNase digestion. As the latter gives AG the the sequence $(U_3C_2)AG$ may be deduced. In experiment 2, where ^{32}P-ATP is used as the input label, both Cp and Gp are observed to be labelled in both the alkaline hydrolysis and the P-RNase digestion. The label is thus *transferred by the method of analysis* from the 5′ side of the input ppp*A label (the asterisk indicates the labelled phosphate) to the 3′-phosphate of both Cp* and Gp*. Thus both C and G must have an A on their 3′-side. As there is only one residue of A and G in the oligonucleotide in question the sequence resolved to (U_3C)-CAG[A]. The square brackets indicate not that this residue is present in the isolated T_1-end product, but that it is the *nearest neighbour*. Experiment 3 demonstrates that ppp*U labels only Up* and not Cp*. The three U's must therefore be consecutive and also on the 5′-side of the C residues. The full sequence is thus UUUCCAG[A]. The information in experiments 4 and 5 is completely consistent with this sequence and serves to confirm it. In summary the sequence of a heptanucleotide may be easily deduced from the results of alkaline hydrolysis alone with the help of the nearest neighbour information. This is in contrast to the methods required for sequencing a *uniformly* labelled heptanucleotide in which partial digestion with venom phosphodiesterase is usually needed. Nevertheless Billeter et al. (1969) had to use the more sophisticated analytical methods described in chs. 6 and 7 for some of the longer oligonucleotides.

The overlapping of the T_1-end products was particularly easy because firstly the residue on the right-hand side of the oligonucleotide was always known from the nearest neighbour information and, secondly, because it was possible to observe the 'time of appearance'

Subject index p. 261

of oligonucleotides by comparing fingerprints after 5, 10, 15, 20, 25 and 30 sec incubations. This information, together with direct isolation of partial digestion products allowed the nearly complete sequence of the first 175 residues to be deduced.

Acknowledgements

I am indebted to a large number of people who have helped me in different ways in the preparation of this monograph. In scientific matters I have freely consulted my colleagues, in particular Drs. Sanger, Smith, Marcker, Barrell and Jeppesen. Dr. Doctor gave invaluable help in the preparation of Ch. 2, which describes techniques many of which are not personally familiar to me. The figures of ch. 2 are reproduced with permission from the following sources: Journal of Biological Chemistry (figs. 2.1, 2.3); Proceedings of the National Academy of Sciences (Washington) (figs. 2.2, 2.4, 2.5); Cold Spring Harbor Symposia on Quantitative Biology (fig. 2.7); Hoppe-Seyler's Zeitschrift für physiologische Chemie (fig. 2.6); Progress in Nucleic Acid Research and Molecular Biology (fig. 2.8 and table 2.2). Figures and plates in other chapters are either from my own work or from that of my colleagues, Drs. Sanger, Barrell, Smith, Labrie, Ambler, Dahlberg and Fellner, to all of whom I am extremely grateful. Much of this is taken from published material and permission has been gratefully received from the relevant Journals: the European Journal of Biochemistry, the Journal of Molecular Biology, and Nature (London).

I am very happy to acknowledge the considerable help of Mrs. Joan Illsley, Mrs. J. Chaplin and Ruth LaTrobe in the long task of typing a manuscript which is as detailed as this. In particular Mrs. Joan Illsley spent many hours in ensuring that the final draft was as accurate, and as free from mistakes, as possible. I am also extremely grateful to Miss Annette Snazle for the careful drawing of the many line diagrams which appear in the manuscript, as well as to Mr. K. Harvey for the preparation of the photographs.

Inevitably mistakes will be present in a manuscript of this length and I would be grateful if readers would inform me of these, or of any other points of criticism. I would then take account of these comments should there be an opportunity for a revised version at any future date.

Appendices

Appendix 1

Tentative rules for representation of nucleosides and polynucleotides*

1.1. One letter symbols for nucleosides

The common ribonucleoside residues (or radicals) are designated by single capital letters, as follows:

A	adenosine
G	guanosine
T	ribosylthymidine
C	cytidine
U	uridine
Ψ	pseudouridine
N	unspecified (or unknown) nucleoside
D**	5,6-dihydrouridine
S	thiouridine

1.1.1. Designation of substituents on bases (rare nucleosides)

The symbols used to indicate modification of a standard base (or nucleoside) are in

---

* An extract of relevant sections from the IUPAC-IUB Commission on Biochemical Nomenclature, 1968-1. The complete rules are available from W. E. Cohn, Office of Biochemical Nomenclature, Oak Ridge National Laboratory, Oak Ridge, Tennessee, U.S.A.

** H is recommended in the rules, but as D is used almost universally, this is likely to be adopted by W. E. Cohn (personal communication) and is therefore used in this book.

lower case letters *immediately before* the single capital letter indicating the standard base (or nucleoside). Some symbols suitable for modification are:

m	methyl
i	isopentenyl
s	thio
ms	methylthio

I.1.2. Designation of substituents on sugars

The symbols are *lower case* if the modified sugar is *internal*, but one may utilise capital letters if the modified sugar occupies a (3'-)terminal position. These symbols are placed *immediately following* (to the right of) the nucleoside symbol, in the usual 3'-position. Without further qualification, this indicates substitution at the (internal) 2'-position. Thus -Am- indicates a 2'-O-methyl-adenosine residue (A^m, or A^m- has been used where space is limiting).

The common, natural termini, phosphate and hydroxyl, are represented, if necessary, by p and oh or ho (many authors use OH or HO); the latter is only required for emphasis as it is implied in the nucleoside symbol itself.

I.1.3. Locants and multipliers

Multipliers, where necessary, are indicated by the usual *sub*scripts: thus -m_2A- signifies a dimethyl adenosine residue, neither methyl being at the 2'-O position. Locants are indicated by *super*scripts: thus -m_2^6A- indicates a 6-dimethyladenosine residue.

I.2. Nucleotides and polynucleotides

I.2.1. Phosphoric acid residues

A monosubstituted (terminal) phosphoric acid is represented by a small p. A phosphoric acid diester (internal) in 3'-5' linkage is represented by a *hyphen* when the sequence is *known*, or by a comma when the sequence is *unknown**. Unknown sequences adjacent to known sequences are placed in parenthesis; these replace, at the points where they occur, the need for other punctuation. All these symbols thus replace the classical 3'-5' or 3'p5' symbols. A 2':3'-cyclic phosphate may be indicated by $>$

Examples: pA-G-A-C(C_2,U_3)A-U-G-C$>$. This sequence has a 5'-phosphate on the A at the 5'-end, a 2':3'-cyclic phosphate on the C at the 3'-end, and a pentanucleotide of

---

* As discussed in § 1.3 the symbol for the phosphodiester bond (p or a hyphen) is omitted in most cases for clarity of representation in this book.

two C's and three U's (sequence unknown) between a C and an A in the middle.

Comments.

The terminal p's* need not be specified if their presence is unknown, in doubt, or of no significance to the argument.

APPENDIX 2

Addresses of suppliers of enzymes, electrophoresis supports and fine chemicals, etc.

Enzymes

1. T_1-RNase
   T_2-RNase } Sankyo Co., Ltd., 7-12 Ginza 2-Chome
   U_2-RNase Chuo-ku, Tokyo, Japan.
2. Bacterial alkaline phosphatase, electrophoretically purified (code BAPF), snake venom phosphodiesterase, pancreatic RNase, polynucleotide phosphorylase, spleen phosphodiesterase (exonuclease), micrococcal nuclease, bovine pancreatic DNase (electrophoretically purified, code DPFF), lysozyme – all from Worthington Biochemical Corp., Freehold, New Jersey, U.S.A.
3. RNase CB (mixture of T_1- and T_2-RNases) from Calbiochem, Ltd., Los Angeles, Calif., U.S.A.

Electrophoresis supports

1. Cellulose acetate. Sheets and strips from Oxoid, Ltd., 20 Southwark Bridge Road, London, S.E.1. Strips from Schleicher and Schüll, 3354 Dassel, W. Germany.
2. Cellogel from Colab Labs., Box 66, Chicago Heights, Ill., U.S.A.
3. Whatman DEAE-Cellulose, Whatman 3 MM and No. 540 papers from H. Reeve, Angel and Co. Ltd., 14 New Bridge St., London, E.C.4.

Column and thin-layer materials

1. Sephadex and DEAE-Sephadex from Pharmacia Fine Chemicals AB, Box 604, Uppsala, Sweden.
2. Chelex resins from Calbiochem, Los Angeles, Calif., U.S.A.
3. Biogels from Bio-Rad Labs., 32nd & Griffin Avenue, Richmond, Calif. 94804, U.S.A.
4. Thin-layer cellulose (MN 300 and MN 300HR) and DEAE-cellulose (MN 300

---

* Some authors use a hyphen for a terminal monophosphate ester as well as for internal phosphodiester linkages. Thus C-C-U-G- is synonymous with C-C-U-Gp.

DEAE) from Macherey, Nagel and Co., D-516, Düren, West Germany (distributed by Camlab (Glass) Ltd., Milton Road, Cambridge, England).

Fine chemicals, etc.

1. BBOT from Ciba, Ltd., Basle, Switserland.
2. CMCT from Fluka, A.G., Buchs, S.G., Switzerland.
3. Acid-washed alumina from Norton Abrasives, Ltd., Welwyn Garden City, Herts, England.
4. Methionine-assay medium from Difco Lab., Detroit 1, Michigan, U.S.A.
5. Marker dyes: acid fuchsin (red), xylene cyanol (blue), orange G (yellow) from Searle Scientific Services, Biological Products, Coronation Rd., High Wycombe, Bucks., England.
6. Macaloid suspension from American Tansul Co., Baroid Division, Nat. Lead Co., 2404 South West Freeway, Houston, Texas, U.S.A.
7. Bovine serum albumin (crystalline) from Sigma Chemical Co. (London), 12 Lettice Street, London, S.W.6.
8. Transfer RNA (*E. coli*) from BDH Chemicals, Poole, Dorset, England. (This is used as a carrier in enzyme digestions. It should be treated with phenol and the RNA precipitated with ethanol as described in § V.1.2. before use.)
9. Yeast nucleic acid from BDH Chemicals, Poole, Dorset, England. (This is used for homochromatography. It may also be used as a carrier for enzyme digestions instead of no. 8 above if treated with phenol.)
10. Millipore filters from Millipore (U.K.) Ltd., Heron House, 109 Wembley Hill Road, Wembley, Middlesex, England.

APPENDIX 3

Addresses of suppliers of equipment

1. High-voltage power packs. Brandenburg, Ltd., 939 London Road, Thornton Heath, Surrey, England; also Locarte, Ltd., 24 Emperors Gate, London, S.W.7.
2. Electrophoresis tanks or parts of tanks. T. W. Wingent, Ltd., 115-117 Cambridge Road, Milton, Cambridge, England. Shandon Southern Instruments Ltd., Camberley, Surrey, England. Savant Instruments, 221 Park Avenue, Hicksville, N.Y. U.S.A. Gilson Medical Electronics, Middleton, Wisconsin, U.S.A.
3. Thin-layer elution device from T. W. Wingent, Ltd., 115-117 Cambridge Road, Milton, Cambridge, England.
4. Fraction collector, 'Uvicord', from LKB-Produkter AB, Box 76, Stockholm, Sweden.
5. Unilux 1 and Unilux 2 bench-top liquid scintillation counters from Nuclear Chicago, Ltd., 333 Howard Avenue, Des Plaines, Ill., U.S.A.

APPENDICES

6. Electric homogenizer from MSE, Ltd., Manor Royal, Crawley, RH10 2ZA, Sussex, England.
7. Portable Geiger counter from Mini-instruments, 36 Southampton Street, London, W.C.2, England.

APPENDIX 4

ε_{mM}* for some nucleotides and nucleosides at pH 7.0 and 260 mμ

Ap	15.0	
Cp	7.6	
Up	10.0	
Gp	11.4	
Ip	7.1	from Holley et al. (1965b)
m^1Gp	11.4	
m^2_2Gp	15.0	
Tp	8.9	
Ψp	9.6	
m_1Ip	7.1	
m^5C	6.0	
m^2G	14	
m^6_2A	14	from Madison et al. (1967a)
m^1A	15	
m^6A	15	
G^m-G	22	

* ε_{mM} for nucleosides and nucleotides are very similar, although not identical (Beaven et al. 1955).

APPENDIX 5

Preparation of radioactively labelled RNA

V.1. Uniformly ^{32}P-labelled E. coli and purification of ^{32}P-rRNA

Several media containing limiting phosphate ion concentrations have been described and are suitable for the preparation of ^{32}P-labelled nucleic acids at high-specific activity (Garen and Levinthal 1960; Gilbert 1963; Hayes et al. 1966; Landy et al. 1967). Although all are satisfactory, we have found that the *L.P. medium* of Landy et al. (1967) gives good yields with incorporation of a high proportion of the added ^{32}P-phosphate.

Low phosphate (L.P.) medium was made from three solutions each sterilised (by autoclaving) separately:

(1) Supplemented salts medium contained 1.5 g KCl, 5.0 g NaCl, 1.0 g NH_4Cl, 2.0 g vitamin-free casamino acids (Difco), 2.0 g bactopeptone (Difco) and 12.1 g Tris in water. The pH was adjusted to 7.4 with concentrated HCl and volume made up to 1 litre with distilled water.

(2) 20% (w/v) glucose.

(3) 0.1M magnesium sulphate.

Completed L.P. medium contained 100 ml of reagent (1), 2 ml of reagent (2) and 1 ml of reagent (3). L.P. Medium (10 ml) was inoculated with *E. coli* (strain CA265 of Dr. S. Brenner or strain MRE600, an RNase-deficient strain (Cammack and Wade 1965) and grown overnight at 37°C in sterile 50-ml test tubes. The medium was aerated by passing compressed air through a glass tube dipping into it to give a fine stream of air bubbles. 5 ml of the resultant suspension of bacteria was inoculated into 100 ml of completed L.P. medium in a 500 ml sterile conical flask containing 25 mC carrier free ^{32}P-labelled phosphate and grown at 37 °C. Adequate aeration is achieved by forcing compressed air through a fine sinter, dipping under the surface of the liquid in the flask so as to produce a fine stream of bubbles. Under these conditions growth is rapid with a doubling time of about 30 min and after 2 hr about 50% of the ^{32}P-phosphate is incorporated into the bacteria. Uptake can be measured by filtration of an aliquot through a Millipore (appendix 2) (cellulose nitrate) filter. The cells are harvested by centrifugation in late exponential phase (uptake between 70 and 90% usually after 2.5-3 hr growth).

V.1.1. *Purification of 16S and 23S rRNA*

To prepare high-molecular weight ribosomal RNA the cell pellet was frozen at −20 °C and then thawed out in the presence of a three-fold excess of acid-washed alumina (appendix 2). The alumina and pellet were moistened with a few drops of a magnesium-containing buffer (0.01 M magnesium acetate, 0.01 M Tris-acetate pH 7.4, 0.006 M mercaptoethanol). The mixture was then ground up with a glass rod at 0 °C for about 5 min and the paste of broken cells was extracted twice with 1 ml portions of the magnesium-containing buffer, the debris and alumina being spun down at 10000 g for 10 min at 0 °C. The supernatant solutions were pooled and centrifuged for 90 min at 105000 g at 0 °C to give a ribosomal pellet freed from DNA and most tRNA. The ribosomal pellet was rinsed with water in the cold, and then dissolved at room temperature in 1 ml of SDS buffer consisting of 0.1 M lithium-chloride, 0.5% (w/v) sodium dodecylsulphate (recrystallised from hot 95% ethanol), 0.001 M EDTA and 0.01 M Tris-chloride, pH 7.5. The solution was loaded onto a 15-30% linear sucrose density gradient (30 ml) prepared with SDS buffer. The gradient was spun for 22 hr at about 10 °C in the SW25 head of the Spinco Model L ultracentrifuge. The fractionated RNA was obtained by collecting drops through a hypodermic needle pushed through the base of the centrifuge tube, and the two main peaks (23s and 16s) were located by counting radioactivity of fractions. A typical count profile is shown in fig. A.1. RNA was recovered by precipitation with 2 volumes of ethanol. A considerable stabilisation against thermal diffusion and agitation of the zones containing the fractionated RNA species is obtained by using gradients

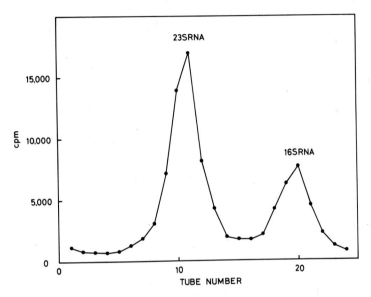

Fig. A.1. Fractionation of ^{32}P-labelled 16S and 23S RNA (*E. coli*) on 15-30% sucrose gradient. For details see text. (From Fellner 1968.)

prepared with such high concentrations of sucrose. The yields are approximately 10% and 20% of the added counts in the 16S and 23S RNA, respectively.

V.1.2. Purification of 5S RNA by electrophoresis on 10% acrylamide slabs

^{32}P-labelled cells (§ V.1) are harvested by centrifugation and resuspended in 1-5 ml of 0.1 M NaCl, 0.01 M Tris-chloride, *p*H 7.5, and shaken vigorously for 1 hr at room temperature with an equal volume of water-saturated re-distilled phenol. The phenol and water layers are separated by centrifugation, and the upper aqueous layer re-extracted with a half volume of fresh phenol before precipitating RNA from the aqueous phase by the addition of two volumes of ethanol. After 1 hr at $-20°$C, crude low-molecular weight RNA is recovered by centrifugation and the ethanol dried in vacuo. 5S RNA can be purified by acrylamide gel electrophoresis where it moves as a sharp band slower

Subject index p. 261

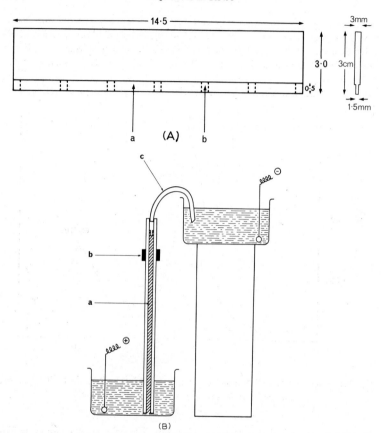

Fig. A.2. Apparatus for electrophoresis on vertical acrylamide gel slabs. (A) Front and side view of a template (or sample slot-former) made of Perspex ($14.5 \times 3.0 \times 0.3$ cm) which is milled down along one of the long edges on either side to form a central projecting tongue (a) approx. 1.5 mm thick. Slots may be cut out of this tongue by filing (e.g. at (b) along the dashed lines) to give any required number of sample areas. For preparative gels it is only necessary to file along the dashed lines at the two edges of the tongue. (B) shows the set-up during electrophoresis. The gel (a) is held between the two thick rectangular glass plates by 'Bulldog' clips and is in electrical contact with the electrode compartments (containing 1 litre of 10-fold diluted 'concentrated' buffer, see text) – the upper one by means of a short paper wick (c) formed from two thicknesses of Whatman No. 3 MM paper. The apparatus is held firmly in position by means of a clamp (b), attached to a retort stand.

than the tRNA region (Hindley 1967). For electrophoresis of ^{32}P-labelled RNA we consider it an advantage to have a simple, versatile and expendable piece of equipment as it often becomes radioactively contaminated; such a piece of equipment developed by the author is described here. A fuller description of the advantages and disadvantages of various types of commercial apparatus appears in Gordon (1969) in volume 1, part I, of this series.

V.1.2.1. An apparatus for vertical gel electrophoresis. The apparatus is shown in fig. A.2. It consists of glass plates sold for thin-layer chromatography ($20 \times 20 \times 0.2$ or $20 \times 40 \times 0.2$ cm); Perspex spacers, 20 (or 40) $\times 2 \times 0.3$ cm and a template as shown in the figure. It is important that the thickness of the Perspex is the same for both spacers and the template and these should, therefore, be cut from the same piece of material. The spacers are greased with Vaseline and inserted to form a seal between each of the long sides of the clean and dry glass plates. The seal is maintained by using 'Bulldog' clips. The apparatus is then inserted vertically into a mould of fresh Plasticine (malleable clay) on the laboratory bench, and pushed down to form a leak-proof seal and held in position using a retort stand. The solution of 10% acrylamide used to form the gel (Peacock and Dingman 1967) consists of the following mixture: 125 ml of a solution of 19% re-crystallised and filtered acrylamide and 1% recrystallised methylene bisacrylamide*, 25 ml of concentrated pH 8.3 buffer (108 g Tris, 9.3 g disodium-EDTA and 55 g of boric acid in 1 litre water), 15 ml of 6.4% dimethylaminopropionitrile and 70 ml water. The solution is warmed to 25 °C and 15 ml of 1.6% ammonium persulphate added. The sol is poured into the apparatus from a beaker, and the template then positioned, avoiding air bubbles, so as to form a slot as near to the top of the gel as possible. The design of the template is such as to form a slot for the sample with the gel on four sides. The dimensions of the slot are alterable (fig. A.2). After 30 min polymerisation is complete and the template may be removed. If the template is rather tightly held, the 'Bulldog' clips at the top may be removed to help loosen it. The top of the gel is then washed with buffer and any loose bits of acrylamide are removed as far as possible. The gel should be examined at this stage, and when held up to the light, no flow lines should be visible except at the top extreme left and right where a certain amount of distortion is unavoidable as some contraction occurs during polymerisation. If leaking occurs before gelation this is invariably due to one of two causes: either a bad seal between glass and Plasticine which is often due to old or rather wet Plasticine; or the template fits too tightly between the glass plates thus prising them apart and weakening the Vaseline seal on each side. This may be rectified by sanding down one of the surfaces of the template with abrasive paper. When leaks do occur the apparatus should be dismantled, cleaned and set up afresh.

---

* Acrylamide may be re-crystallised from chloroform (50 g/l) at 50 °C or from ethyl acetate (400 g/l) at 50 °C, cooled to -20 °C and filtered. Methylene bisacrylamide may be similarly re-crystallised from acetone (10 g/l).

Attempts at adding fresh solution of acrylamide to a slowly leaking gel always result in poorly formed gels and bad runs.

The apparatus containing the gel is then set up in a cold room at 4 °C as shown in fig. A.2. The upper end of the gel is connected by a paper wick (a double thickness of Whatman No. 3 MM paper) to an upper electrode compartment. The wick must be as short as possible to avoid uneven conductivity which results in distortion of the bands during the run. The level of liquid in the upper reservoir (holding approximately 1 litre) should be 1 cm higher than the level of liquid in the glass, to avoid drying out.

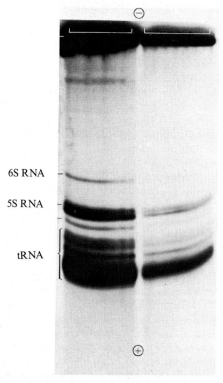

Fig. A.3. Radioautograph of a low molecular weight RNA preparation of *E. coli* (Q13) fractionated on a 10% acrylamide gel slab. The two samples were extracted under slightly different conditions. On the left is a standard RNA preparation (§ V.1.2), while on the right is a hot phenol-SDS extraction (Warner et al. 1966) of the residual phenol and interphase material of the first phenol extraction (§ V.1.2). The pattern of bands appears similar in both samples. Notice the resolution of 5S RNA into two components, and of tRNA into a major and several slower components.

The gel is pre-run at 350 V (over 40 cm) for 2-3 hr to remove ionic impurities and then the sample loaded with a fine drawn-out glass tube inserted into the slot. Up to 0.4 ml and 2-3 mg may be loaded on the full width of gel with good results. Larger volumes may be loaded with some loss in resolution particularly of faster moving bands. Ideally, however, the lowest possible volume should be used. The sample should not be so large as to spill over into the liquid above the slot. The sample should be soluble (insoluble material being spun off prior to loading), be protein-free and should be in lower (usually half the) concentration of the same buffer as is used in the gel. It should also contain some bromphenol blue and 20% sucrose. Loading of the sample is difficult without the blue colour and the sucrose additives. The length of run will depend on the separation required but for the fractionation shown in fig. A.3 it was 18 hr at 350 V over 40 cm. The amperage was at first 30 mA and dropped to about 15 mA at the end of the run.

The radioautograph is prepared by prising off one glass plate and laying 'Saranrap' (thin cellophane sandwich wrapping) on the surface of the gel. A strip of linen tape marked at intervals with ^{32}P-phosphate-containing red ink is taped down on top at each edge and a radioautograph is prepared by exposing for usually 1-30 min. The radioautograph is aligned with the ink marks and bands cut out using the radioautograph (itself cut-out) as a guide. It is convenient to use a scalpel for this, covering that part of the gel not being cut with a glass plate so as to avoid unnecessary exposure to irradiation from the ^{32}P.

V.1.2.2. Elution. The following procedure has been used for 5S RNA. The gel is homogenized with approximately 5 ml of 1.0 M NaCl using a Potter (or motor-driven) homogeniser. The use of a hand homogeniser is not sufficient to break up the rather rigid gel. The suspension is left at 0 °C for 30 min and the supernatant recovered by centrifugation. The gel sediment is homogenised and extracted twice more with NaCl. The combined supernatant fractions are then diluted to 0.3 M NaCl with water and loaded onto a 1 × 1 cm internal diameter DEAE-cellulose column equilibrated with 0.3 M NaCl, 0.01 M Tris-chloride, *p*H 7.5 at room temperature. The column is washed with 10 ml of buffer and the RNA eluted with about 3 ml of 1.0 M NaCl in 7 M urea. After adding 50-100 µg of carrier tRNA (appendix 2) ^{32}P-labelled 5S RNA is precipitated by adding 2 volumes of ethanol and recovered by centrifugation and dried in vacuo. If salt (whitish) is still visible, it is dissolved in 0.1 M NaCl and RNA re-precipitated and recovered as above. The yield was approximately 0.1% of the radioactivity added to the growth medium and the specific activity was approximately 2×10^6 dpm/µg.

V.2. Uniformly ^{32}P-labelled yeast RNA

The medium of Williams and Dawson (1952), modified by the replacement of phosphate salts by ammonium citrate at *p*H 5.2 gives suitable incorporation of carrier-free ^{32}P-phosphate.

1 litre of the medium contained the following: glucose 20 g; triammonium citrate,

pH 5.2, 3.25 g; $MgSO_4 \cdot 7H_2O$, 0.25 g; sodium citrate 1.0 g; L-asparagine monohydrate 2.5 g; biotin, 10 μg; calcium pantothenate 0.5 mg; inositol, 10 mg; thiamine 6 mg; pyridoxine 1 mg; zinc acetate, 400 μg; ferric chloride 150 μg, copper chloride 25 μg. It was convenient to prepare three separate mixtures of the salts, the vitamins and the heavy metals solution and to sterilise by using an autoclave before mixing in the correct proportions to make up the completed medium. 10 mC of ^{32}P-phosphate and 1-5 mg of fresh bakers' yeast were added to 50 ml of medium which was incubated in a 250 ml conical flask at 30 °C for 6 hr and aerated by passing compressed air through a tube dipping under the surface of the medium to produce a fine stream of bubbles. The uptake of radioactivity into the yeast was 95% complete in this time. Cells were collected by centrifugation. A low molecular weight RNA preparation may be made as described in §V.1.2 above. High-molecular weight rRNA is extracted as in §V.1.1.

V.3. Uniformly ^{32}P-labelled mammalian RNA

Mammalian cells such as human KB cells (human epidermoid carcinoma line) and Landschutz (mouse tumour line of ascitic origin) grow well in suspension culture. Forget and Weissman (1967) obtained ^{32}P-labelled 5S RNA of high specific activity by labelling in phosphate-free Eagle's medium (Eagle 1959), supplemented with dialysed 5% unheated horse serum, 2 mM glutamine, 10^{-5} M phosphate and carrier-free ^{32}P-phosphate. Cells were grown at approximately 2×10^5 cells/ml and 10 μC of ^{32}P-phosphate/ml for 24-48 hr (generation time of 24 hr). Williamson and Brownlee (1969) used instead Waymouth's medium containing low phosphate Hank's balanced salt solutions (Paul 1965). For further details of the rather specialised techniques and media required for tissue culture of mammalian cells, reference should be made to the authors using the particular cell lines.

V.4. Uniformly ^{32}P-labelled bacteriophage R17 (f2 or Qβ) RNA
(modified by Dr. J. Steitz from Lodish et al. 1965)

Purification of small amounts of high-specific activity phage was accomplished as follows. 2 ml of an overnight culture of *E. coli* (strain carrying an F episome) was inoculated into 200 ml of the following sterile medium. 40 ml Peptone (5% w/v bactopeptone – Difco, in water), 1 ml 20% w/v glucose, 5 ml 20% w/v NaCl, 154 ml distilled water. Cells were grown at 37 °C in a 500-ml Roux bottle, aerated through a fine glass sinter until the optical density of the cell suspension was 0.5-0.6 at 550 mμ (about 5×10^8 cells/ml). 0.4 ml of 1 M $CaCl_2$, phage at 10 plaque-forming units/cell (i.e. 5×10^9/ml) and 20 mC carrier-free ^{32}P-phosphate were then added and growth was continued for a further 6-8 hr. It was important to aerate rather gently for the first 5-10 min of phage infection while 'phage attachment' occurred. At the end of growth, lysis of cells was completed by the addition of 3-4 drops of chloroform, lysozyme (appendix 2) to a final concentration of 0.1 mg/ml and 5 ml of 0.1 M EDTA. Aeration was continued for a further 15 min.

Phage was assayed at this stage by plating 10-fold serial dilutions onto bacterial lawns on agar plates. After incubating overnight at 37 °C, plaques were counted. The yield should be about 10^{12} p.f.u. per ml.

Phage was precipitated from the medium by the addition of 80 ml of uninfected, unlabelled *E. coli* (added as carrier) and 0.3 g $(NH_4)_2SO_4$ per ml of medium and left overnight at 0 °C. Phage was recovered by centrifugation for 20 min at 40000 g. The precipitate was resuspended in a small volume of SSC (0.15 M NaCl, 0.015 M sodium citrate, pH 7.0), and transferred to a glass homogeniser with a close fitting Teflon plunger. 25 μl of DNase solution at 2 mg/ml (appendix 2) and 200 μl of 0.1 M $MgCl_2$ was added and the volume made up to 20 ml with the SSC. The suspension was kept in the cold and homogenised at frequent intervals over a period of 2-3 hr. Cell debris was removed by centrifugation for 15 min at 40000 g and the phage was recovered from the supernatant by pelleting in the 40 rotor of the Spinco Model L ultracentrifuge for 3 hr at 40000 r.p.m. The phage pellet was resuspended in 1.0 to 1.5 ml SSC, 2.6 g of $CsCl_2$ was added and the volume made up to 4.5 ml with distilled water for centrifugation for 20 hr at 39000 r.p.m. in the S.W. 39 rotor of the Spinco Model L. Phage is banded at its buoyant density and was located as the major radioactive peak by collecting fractions and counting. The material in the phage peak was pooled, diluted with at least six times its volume of distilled water and the RNA was extracted by shaking vigorously with water-saturated phenol for 30 min in the cold. After separating the aqueous layer by centrifugation, the phenol and interface material were re-extracted twice more with water. The combined aqueous layers were briefly shaken again with phenol and the RNA was precipitated from the aqueous layer by the addition of 2.5 vols of ethanol. RNA was recovered by centrifugation and the precipitate dried in vacuo. The yield was 1-2 mC (5-10%) and the specific activity was approximately 1 mC per mg of RNA.

V.5. (^{14}C-methyl and ^3H-methyl)-labelled RNA of E. coli
(Fellner 1968)

0.1-1 mC of (^{14}C-methyl)-methionine of approximately 50 mC/mmole (or 5 mC of (^3H-methyl) methionine (5 C/mmole) is used to label *E. coli*. A methionine requiring auxotroph, for example CB3 of Dr. S. Brenner (K12 W6, met$^-$, leu$^-$, thr$^-$) is used to obtain the highest possible incorporation. The strain must be grown in the presence of methionine, leucine and threonine to ensure normal methylation of RNA (Mandel and Borek 1961 and 1963). A methionine assay medium (appendix 2) was used for growth of the bacteria. The solid nutrient was dissolved and sterilised according to the manufacturers' instructions. The resultant solution was then diluted ten times with 0.1 M Trischloride, pH 7.5, and was used, with the addition of 0.01 mg/ml of L-methionine, for growing an inoculum overnight. The inoculum was then added to a larger volume of fresh methionine-assay medium containing the (^{14}C-methyl)-methionine at a final concentration of about 0.01 mg/ml. Growth was carried out at 37 °C in a 250 ml conical flask aerated through a fine glass sinter. After 6 hr (late logarithmic phase) the bacteria

Subject index p. 261

were harvested, when about 90% of the (^{14}C-methyl)-methionine had been incorporated into the cells. RNA may be extracted and purified by methods already described (§§ V.1.1 and V.1.2).

V.6. ^{35}S-labelled E. coli

E. coli was grown up overnight on 100 ml of the following medium (which contained per litre: 5.8 g Na_2HPO_4, 3.0 g KH_2PO_4, 0.5 g NaCl, 1.0 g NH_4Cl, 0.05% w/v glucose and 10^{-3} M $MgSO_4$) at 37 °C with efficient aeration, as above. In the morning, the cells were harvested by centrifugation and resuspended in fresh medium containing $MgCl_2$ instead of $MgSO_4$ and 0.4% glucose. After 15 min growth, 5-20 mC of carrier-free ^{35}S-sulphate was added. Uptake of radioactivity was usually quantitative after 2-4 hr and the cells were harvested by centrifugation.

APPENDIX 6

Sequence of some oligonucleotides from rRNA (see fig. 4.9)

Spot 46 of fig. 4.9a	AACCG
Spot 61 of fig. 4.9a	UACCG
Spot 63 of fig. 4.9a	AUCCG
Spot 89 of fig. 4.9a	UUCCG
Spot 66 of fig. 4.9b	GAAAGC

APPENDIX 7

Summary of conditions of enzyme digestion and of fractionation systems for radioactive sequencing

VII.1. Initial digestion of RNA

In order to use the paper fractionation techniques it is necessary to digest RNA in fairly small volumes. The enzymatic digestions are conveniently done as follows: A sample of ^{32}P-RNA is dried down in vacuo in a small silicone-treated tube and then dissolved in the appropriate amount of enzyme, buffer and carrier tRNA (if necessary) in a total volume of not more than 5 μl and not more than 50 μg RNA. The incubation is then carried out in the tip of a drawn-out capillary tube and the mixture is then loaded directly onto the cellulose acetate for fractionation in the first dimension of the two-dimensional system.

VII.1.1. Complete digestion with T_1-RNase

The RNA is digested for $\frac{1}{2}$ hr at 37 °C, using about 1/10 the weight of enzyme to RNA, in 0.01 M Tris-chloride, 0.001 M EDTA, pH 7.4.

VII.1.2. Complete digestion with pancreatic RNase

The conditions are the same as for T_1-RNase above, except that about 1/20 the weight of enzyme to substrate is used.

VII.1.3. Combined T_1- and alkaline phosphatase digest

This is an important type of degradation as it liberates oligonucleotides which lack a 3'-terminal phosphate. These products are then susceptible to further partial digestion using snake venom phosphodiesterase. Incubation is for 60 min at 37 °C using an enzyme to substrate ratio of 1 : 10 for T_1-RNase and 1 : 5 for bacterial alkaline phosphatase in 0.01 M Tris-chloride, pH 8.0.

VII.1.4. Partial digestion with either T_1- or pancreatic RNases

The conditions of enzyme digestion depend on the extent of degradation required. However suitable average conditions are enzyme to substrate ratios of between 1 : 100 to 1 : 500 in 0.01 M $MgCl_2$, 0.01 M Tris-chloride, pH 7.5, and digesting for about 30 min at 0 °C (pre-cool sample and solutions).

VII.2. Two-dimensional fractionation systems

VII.2.1. The standard two-dimensional ionophoretic system is used for studying the small oligonucleotides up to about 10 residues long, (the limiting size depends on U content). This method uses ionophoresis on cellulose acetate strips at pH 3.5 in 7 M urea, after which the oligonucleotides are transferred by 'blotting' onto DEAE-paper and fractionated at right-angles to the first dimension using 7% formic acid.

VII.2.2. A modified two-dimensional system using the same technique for the first dimension but 'homochromatography' for the second. The homochromatography is carried out on DEAE thin layers. They are run at 60 °C in an oven and provide a rather versatile system for fractionating oligonucleotides from approximately 10-50 residues in length, depending on the composition of the 'homomixture' used. The thin layers are prepared from a 1 : 7.5 or a 1 : 10 mixture of DEAE-cellulose to cellulose on long glass plates. Mix *a* is a 3-5% mixture of yeast nucleic acid in 7 M urea. Mix *b* is a 3-5% mixture, dialysed for 4 hr to remove salt, in 7 M urea. Mix *c* is a 3-5% mix hydrolysed for (usually) between 5 and 30 min (depending on the fractionation required) in 1 N KOH at room temperature before neutralising, dialysing and making 7 M with respect to urea. After drying off the thin layer, it is marked with red ink containing ^{35}S-labelled sulphate, and then placed in contact with an X-ray film in a lead-lined folder for autoradiography.

VII.3. Sequence of oligonucleotides

Although a certain amount of information may be deduced by comparison with known 'maps' of spots, all spots should be eluted and further characterised. Elution is with 30% triethylamine carbonate, pH 10 (containing carrier RNA at 0.4 mg/ml except when eluting from thin layers). The detailed procedure depends on which type of fractionation system has been used. Digestion is normally carried out in about 10 μl volumes in sealed capillary tubes. For the shorter incubations (30 min) it is not necessary to seal the tubes if they are kept in a horizontal position during the incubation. The digestion is stopped by loading directly onto the paper without desalting. Samples may be conveniently stored on paper if electrophoresis tanks are not immediately available for fractionation. Alternatively, after digestion in capillary tubes, these may be frozen at $-20\,°C$ before loading onto the paper.

VII.3.1. Composition

The spot (freed from triethylamine carbonate) is dissolved in about 10 μl of 0.2 N NaOH and incubated at 37 °C for about 16 hr and then fractionated by ionophoresis on Whatman No. 540 paper at pH 3.5. This system conveniently separates the four mononucleotides. For nucleotides from thin layers 0.5 N NaOH and No. 3 MM paper are used.

VII.3.2. Digestion of T_1-oligonucleotides with pancreatic RNase or pancreatic oligonucleotides with T_1-RNase

Digestion in both cases is with a solution of 0.1 mg/ml of enzyme, 2 mg/ml carrier RNA* in 0.01 M Tris-chloride, 0.001 M EDTA, pH 7.4 for 30 min at 37 °C. Fractionation may be carried out by DEAE-paper ionophoresis at pH 3.5. For nucleotides from thin layers digestion is for 2 hr and the carrier RNA is omitted.

VII.3.3. Digestion of T_1-oligonucleotides with U_2-RNase (A specific)

Digestion is with 10 μl of between 0.1 and 1 unit/ml in 0.05 M sodium acetate, 0.002 M EDTA, pH 4.5 containing 0.1 mg/ml bovine serum albumin for 2 hr at 37 °C. For

---

* The carrier may be omitted if already introduced in sufficient concentration during the elution procedure (§ VII.3 above).

nucleotides from thin layers 10 units/ml of enzyme is used and incubation is 4 hr at 37 °C. Fractionation is by DEAE-paper ionophoresis using the pH 1.9 system.

VII.3.4. Cleavage of T_1-oligonucleotides next to C residues by carbodiimide blocking

10 μl of the reagent (CMCT) at 20-50 mg/ml (100 mg/ml for spots from thin layers) in 0.05 M Tris-chloride, 0.005 M EDTA, pH 7.5, is incubated with the nucleotide overnight at 37 °C. 10 μl of pancreatic RNase (0.2 mg/ml) is added and incubation continued for a further half to one hour at 37 °C (1-2 hr for spots from thin layers). Fractionation is on 3 MM paper, placing the origin in the middle of the paper at pH 3.5. Modified nucleotides are basic and may be analysed by alkaline hydrolysis, or by unblocking with ammonia (0.2 N ammonia overnight at room temperature) and further digestion with pancreatic RNase.

VII.3.5. 3'- and 5'-end groups

Digestion is with 0.2 mg/ml venom phosphodiesterase in 0.05 M Tris-chloride, 0.01 M magnesium chloride, pH 8.9 for 2 hr at 37 °C and fractionation for $1\frac{1}{2}$ hr at 3 kV on No. 540 paper at pH 3.5. The 5'-end is liberated as a nucleoside and can be identified by comparison with the alkaline digest. The 3'-end is liberated as a mononucleoside 3', 5'-diphosphate. Diphosphates run faster than the monophosphates.

VII.3.6. Analysis of T_1-oligonucleotides by partial digestion with spleen exonuclease

Suitable conditions are 0.2 mg/ml of enzyme in buffer (0.1 M ammonium acetate, pH 5.7, 0.002 M EDTA, 0.05% Tween 80) digesting for 0 min (a control), 30 min and 1 hr at room temperature and fractionating on DEAE-paper at pH 1.9.

VII.3.7. Analysis of pancreatic oligonucleotides as for § VII.3.6

Lower concentrations of enzymes are required and 0.1 mg/ml is suitable.

VII.3.8. Analysis of oligonucleotides lacking a 3'-terminal phosphate group by partial digestion with snake venom phosphodiesterase (for products from T_1 and phosphatase digests).

Spots are digested with 0.01 mg/ml enzyme in buffer (0.05 M Tris-chloride, pH 8.9, 0.01 M $MgCl_2$) for 0 min (a control), 10, 20 and 30 min at 37 °C and fractionated by ionophoresis on DEAE-paper at pH 3.5. Products must be eluted and their composition determined by alkaline hydrolysis.

VII.3.9. Complete digestion of oligonucleotides lacking a 3'-terminal phosphate group with snake venom phosphodiesterase

Digestion is for 1 hr at 37 °C using 0.1 mg/ml of enzyme in buffer and carrier RNA* (0.01 M Tris-chloride, 0.01 M magnesium acetate (or chloride), pH 9.0 and 1 mg/ml carrier RNA) and fractionation by paper ionophoresis on No. 540 paper at pH 3.5.

VII.3.10. Analysis of partial digestion products

Usually the eluted spot will be divided into two, and half analysed by further digestion with T_1-RNase and the other half with pancreatic RNase. The T_1-RNase digest may be fractionated by ionophoresis on DEAE-cellulose at 7% formic acid, and the pancreatic RNase digest at pH 1.9 or 7% formic acid. Products may be tentatively identified from their position and confirmed by elution and further analysis. The concentration of enzyme necessary for this further enzymatic digest is high because of the carrier oligonucleotides introduced by homochromatography on the thin layers. A solution of 0.1 mg/ml T_1-RNase in 0.01 M Tris-chloride, 0.001 M EDTA, pH 7.5, is suitable and digestion is for 1 hr at 37 °C. Exactly the same conditions are used for pancreatic RNase.

VII.3.11. T_2-RNase digestion

Digestion is carried out with 2 units of enzyme/ml in 0.05 M ammonium acetate, pH 4.5 containing 0.1 mg/ml bovine serum albumin for 2 hr at 37 °C. Fractionate as in § VII.3.1. above. T_1-RNase and P-RNase may be added, each at a concentration of 0.05 mg of enzyme per ml, to the above mixture.

VII.4. Solutions

The following solutions are required for the methods outlined above. Solutions are made up in distilled water and are stored frozen at $-20\,°C$ (with the exception of items 8, 9, 14 and 15) in order to prevent bacterial growth.

1. P-RNase, 2 mg/ml.
2. T_1-RNase, 2 mg/ml.
3. 0.1 M Tris-chloride, 0.01 M EDTA, pH 7.4.
4. 0.02 M Tris-chloride, 0.02 M $MgCl_2$, pH 7.4.
5. 0.1 M Tris-chloride, pH 8.0.
6. 12.5 mg/ml bacterial alkaline phosphate, freshly made up from the ammonium

---

* The carrier may be omitted if already introduced in sufficient concentration during the elution procedure (§ VII.3 above).

sulphate precipitate supplied by centrifuging a known volume of suspension and redissolving precipitate in water.
7. Carrier tRNA, 50 mg/ml.
8. 30% TEC, pH 10, with carrier (0.4 mg/ml).
9. 30% TEC, pH 10, without carrier.
   Items 8 and 9 are kept at 4°; their pH is checked before use and adjusted, if necessary, with triethylamine.
10. 0.5 M Tris-chloride, 0.1 M $MgCl_2$, pH 8.9.
11. 0.01 M Tris-chloride, 0.001 M EDTA, pH 8.9.
12. 0.05 M Na-acetate, 0.002 M EDTA, pH 4.5 + 10 units/ml U_2-RNase + 0.1 mg/ml BSA.
13. 0.05 M Na-acetate, 0.002 M EDTA, pH 4.5 + 1.0 units/ml U_2-RNase + 0.1 mg/ml BSA.
14. 0.5 N NaOH ⎫ stored at room temperature.
15. 0.2 N NaOH ⎭
16. 0.1 mg/ml T_1-RNase in 0.01 M Tris-chloride, 0.001 M EDTA, pH 7.4.
17. 0.1 mg/ml P-RNase in 0.01 M Tris-chloride, 0.001 M EDTA, pH 7.4.
18. 0.1 mg/ml T_1-RNase in 0.01 M Tris, 0.001 M EDTA, pH 7.4, 2.0 mg/ml carrier RNA.
19. 0.1 mg/ml P-RNase in 0.01 M Tris, 0.001 M EDTA, pH 7.4, 2.0 mg/ml carrier RNA.
20. 0.2 mg/ml P-RNase in 0.1 M Tris-chloride, 0.01 M EDTA, pH 7.4.
21. 0.2 mg/ml venom phosphodiesterase in 0.05 M Tris-chloride, 0.01 M $MgCl_2$, pH 8.9.
22. 0.1 mg/ml venom phosphodiesterase in 0.01 M Tris-chloride, 0.01 M $MgCl_2$ and 2 mg/ml carrier RNA, pH 9.0.
23. 0.2 mg/ml spleen phosphodiesterase in 0.1 M ammonium acetate, 0.002 M EDTA, 0.05% Tween 80, pH 5.7.
24. 0.1 mg/ml spleen phosphodiesterase in same buffer as in 23.
25. 0.1 mg/ml venom phosphodiesterase.

References

ADAMS, J. M., P. E. N. JEPPESEN, F. SANGER and B. G. BARRELL (1969) Nature (Lond.) *223*, 1009

APGAR, J., G. A. EVERETT and R. W. HOLLEY (1965) Proc. Natl. Acad. Sci. (Wash.) *53*, 546

APGAR, J., G. A. EVERETT and R. W. HOLLEY (1966) J. Biol. Chem. *241*, 1206

ARIMA, T., T. UCHIDA and F. EGAMI (1968) Biochem. J. *106*, 609

AUBERT, M., J. F. SCOTT, M. REYNIER and R. MONIER (1968) Proc. Natl. Acad. Sci. (Wash.) *61*, 292

AUGUSTI-TOCCO, G. and G. L. BROWN (1965) Nature (Lond.) *206*, 683

AVERY, O. T., C. M. MACLEOD and M. MCCARTY (1944) J. Exptl. Med. *79*, 137

BACZYNSKYJ, L., K. BIEMANN and R. W. HALL (1968) Science *159*, 1481

BAEV, A. A., T. V. VENKSTERN, A. D. MIRZABEKOV' A. I. KRUTILINA, L. LI and V. D. AKSEL'ROD (1967) Mol. Biol. USSR *1*, 754

BARRELL, B. G. and F. SANGER (1969) FEBS Letters *3*, 275

BEAVEN, G. H., E. R. HOLIDAY and E. A. JOHNSON (1955) The nucleic acids, Vol. I, p. 493 (Academic Press, New York)

BERNARDI, A. and G. BERNARDI (1966) Biochim. Biophys. Acta *129*, 23

BILLETER, M. A., J. E. DAHLBERG, H. M. GOODMAN, J. HINDLEY and C. WEISSMAN (1969) Nature (Lond.) *224*, 1083

BOCK, R. M. (1967) Methods in enzymology, Section 29, Vol. 12, Nucleic acids, Part A (Academic Press, New York and London)

BOEDTKER, H. and D. G. KELLING (1967) Biochem. Biophys. Res. Commun. *29*, 758

BRIMACOMBE, R. L. C., B. E. GRIFFIN, J. A. HAINES, W. J. HASLAM and C. M. REESE (1965) Biochemistry *4*, 2452

BROWN, D. M. and A. R. TODD (1952) J. Chem. Soc. 52

BROWNLEE, G. G. (1971) Nature, *229*, 147

BROWNLEE, G. G. and F. SANGER (1967) J. Mol. Biol. *23*, 337

BROWNLEE, G. G. and F. SANGER (1969) European J. Biochem. (1969) *11*, 395

BROWNLEE, G. G., F. SANGER and B. G. BARRELL (1967) Nature (Lond.) *215*, 735
BROWNLEE, G. G., F. SANGER and B. G. BARRELL (1968) J. Mol. Biol. *34*, 379
BURROWS, W. J., D. J. ARMSTRONG, F. SKOOG, S. M. HECHT, J. T. A. BOYLE, N. J. LEONARD and J. OCCOLOWITZ (1968) Science *161*, 691
CAMMACK, K. A. and H. E. WADE (1965) Biochem. J. *96*, 671
CANTOR, C. R. (1967) Nature (Lond.) *216*, 513
CARBON, J., H. DAVID and M. H. STUDIER (1968) Science, *161*, 1147
CHANG, S. H. and U. L. RAJBHANDARY (1968) J. Biol. Chem. *243*, 592
CORY, S. K. and K. A. MARCKER (1970) European J. Biochem. *12*, 177
DAHLBERG, J. E. (1968) Nature (Lond,) *220*, 548
DE WACHTER, R. and W. FIERS (1967) J. Mol. Biol. *30*, 507
DE WACHTER, R., J.-P. VERHASSEL and W. FIERS (1968a) Biochim. Biophys. Acta *157*, 195
DE WACHTER, R., J.-P. VERHASSEL and W. FIERS (1968b) FEBS Letters *1*, 93
DUBE, S. K. and K. A. MARCKER (1969) European J. Biochem. *8*, 256
DUBE, S. K., K. A. MARCKER, B. F. C. CLARK and S. CORY (1969) European J. Biochem. *8*, 244
DUBE, S. K., K. A. MARCKER and A. YUDELEVICH (1970) FEBS Letters *9*, 168
DUNN, D. B. (1963) Biochem. J. *86*, 14P
DUTTING, D., H. FELDMANN and H. G. ZACHAU (1966) Hoppe-Seylers Z. Physiol. Chem. *347*, 249
EAGLE, H. (1959) Science *130*, 432
FELDMANN, H., D. DUTTING and H. G. ZACHAU (1966) Hoppe-Seylers Z. Physiol. Chem. *347*, 236
FELLNER, P. (1968) Ph. D. Thesis, Cambridge University
FELLNER, P. (1969) European J. Biochem. *11*, 12
FELLNER, P., C. EHRESMANN and J. P. EBEL (1970) Nature (Lond.) *225*, 26
FELLNER, P. and F. SANGER (1968) Nature (Lond.) *219*, 236
FISCHER, L. (1969). Laboratory techniques in biochemistry and molecular biology, Vol. 1 (North-Holland, Amsterdam-London)
FORGET, B. G. and S. M. WEISSMAN (1967) Nature (Lond.) *213*, 878
FORGET, B. G. and S. M. WEISSMAN (1968) J. Biol. Chem. *243*, 5709
GAREN, A. and C. LEVINTHAL (1960). Biochim Biophys. Acta *38*, 470
GILBERT, W. (1963) J. Mol. Biol. *6*, 389
GILHAM, P. T. (1962) J. Am. Chem. Soc. *84*, 687
GILLAM, I., D. BLEW, R. C. WARRINGTON, M. VON TIGERSTROM and G. M. TENER (1968) Biochemistry 7, 3459
GLYNN, I. M. and J. B. CHAPPELL (1964) Biochem. J. *90*, 147
GOODMAN, H. M., J. N. ABELSON, A. LANDY, S. ZADRAZIL and J. D. SMITH (1970) European J. Biochem. *13*, 461
GORDON, A. H. (1969) Laboratory techniques in biochemistry and molecular biology, Vol. 1 (North-Holland, Amsterdam-London)

Gray, M. W. and B. G. Lane (1968) Biochemistry 7, 3441
Gross, D. (1961) J. Chromatog. 5, 194
Hall, R. H. (1963a) Biochem. Biophys. Res. Commun. 13, 394
Hall, R. H. (1863b) Biochem. Res. Commun. 5, 361
Hall, R. H. (1965) Biochemistry 4, 661
Hall, R. H., L. Csonka, H. David and B. McLennan (1967) Science 156, 69
Harada, F., H. J. Gross, F. Kimura, S. H. Chang, S. Nishimura and U. L. Raj-Bhandary (1968) Biochem. Biophys. Res. Commun. 33, 299
Hayes, D. H., F. Hayes and M. F. Guerin (1966) J. Mol. Biol. 18, 499
Heppel, L. A. (1967) Methods in enzymology, Vol. 12, Nucleic acids, p. 316 and 317 (Academic Press, New York and London)
Hindley, J. (1967) J. Mol. Biol. 30, 125
Hirsh, D. (1971) J. Mol. Biol. 58, 439
Hiramaru, M., T. Uchida and F. Egami (1966) Anal. Biochem. 17, 135
Holley, R. W. (1968) Progress in nucleic acid research and molecular biology, Vol. 8 (Academic Press, New York and London)
Holley, R. W., J. Apgar, B. P. Doctor, J. Farrow, M. A. Marini and S. H. Merrill (1961) J. Biol. Chem. 236, 200
Holley, R. W., J. Apgar, G. A. Everett, J. T. Madison, M. Marquisee, S. H. Merrill, J. R. Penswick and A. Zamir (1965a) Science 147, 1462
Holley, R. W., J. Apgar, G. A. Everett, J. T. Madison, S. H. Merrill and A. Zamir (1963) Cold Spring Harbour Symp. Quant. Biol. 28, 117
Holley, R. W., G. A. Everett, J. T. Madison and A. Zamir (1965b) J. Biol. Chem. 240, 2122
Holley, R. W., J. T. Madison and A. Zamir (1964) Biochem. Biophys. Res. Commun. 17, 389
Hurwitz, J., J. J. Furth, M. Anders, P. J. Ortiz and J. T. August (1961) Cold Spring Harbour Symp. Quant. Biol. 26, 91
Jeppesen, P. G. N. (1971) Biochem J. 124, 357
Katz, A. M., W. J. Dreyer and C. B. Anfinsen (1959) J. Biol. Chem. 234, 2897
Keller, E. B. (1964) Biochem. Biophys. Res. Commun. 17, 412
Labrie, F. and F. Sanger (1969) Biochem. J. 114, 29P
Landy, A., J. Abelson, H. M. Goodman and J. D. Smith (1967) J. Mol. Biol. 29, 457
Lawley, P. D. and P. Brookes (1963) Biochem. J. 89, 127
Lawley, P. D., and C. A. Wallick (1957) Chem. Ind. p. 633
Lee, J. C., N. W. Y. Ho and P. T. Gilham (1965) Biochim. Biophys. Acta 95, 503
Lee, J. C. and V. M. Ingram (1969) J. Mol. Biol. 41, 431
Lehman, I. R., M. J. Bessman, E. S. Simms and A. Kornberg (1958) J. Biol. Chem. 233, 163
Linn, S. and I. R. Lehman (1965) J. Biol. Chem. 240, 1287
Lipsett, M. N. (1965) J. Biol. Chem. 240, 3975
Lipsett, M. N. and B. P. Doctor (1967) J. Biol. Chem. 242, 4072

LODISH, H. F., K. HORIUCHI and N. P. ZINDER (1965) Virology 27, 139
MADISON, J. T., G. A. EVERETT and H. KUNG (1966) Science 153, 531
MADISON, J. T., G. A. EVERETT and H. KUNG (1967a) J. Biol. Chem. 242, 1318
MADISON, J. T. and R. W. HOLLEY (1965) Biochem. Biophys. Res. Commun. 18, 153
MADISON, J. T., R. W. HOLLEY, J. S. POUCHER and P. H. CONNETT (1967b) Biochim. Biophys. Acta 145, 825
MADISON, J. T. and H. KUNG (1967) J. Biol. Chem. 242, 1324
MANDEL, L. R. and E. BOREK (1961) Biochem. Biophys. Res. Commun. 6, 138
MANDEL, L. R. and E. BOREK (1963) Biochemistry 2, 555
MARKHAM, R. (1957) Methods in enzymology, Vol. 3 (Academic Press, New York)
MARKHAM, R. (1963) Prog. Nucleic Acid Res. 2, 61
MARKHAM, R. and J. D. SMITH (1952) Biochem. J. 52, 552
MATTHEWS, H. and H. GOULD, in prep. Laboratory techniques in biochemistry and molecular biology, Vol. 4 (North-Holland, Amsterdam-London)
MICHL, H. (1951) Monatsh. Chem. 82, 489
MICHL, H. (1952) Monatsh. Chem. 83, 737
MIN-JOU, W. and W. FIERS (1969). J. Mol. Biol. 40, 187
MORELL, P., I. SMITH, D. DUBNAU and J. MARMUR (1967). Biochemistry 6, 258
MURAO, K., M. SANEYOSHI, F. HARADA and S. NISHIMURA (1970). Biochem. Biophys. Res. Commun. 38, 657
NAKANISHI, K., N. FURUTACHI, M. FUNAMIZU, D. GRUNBERGER and I. B. WEINSTEIN (1970). J. Am. Chem. Soc. 92, 7617
NAUGHTON, M. A. and H. HAGOPIAN (1962). Anal. Biochem. 3, 276
NICHOLS, J. L. and B. G. LANE (1966). Biochim. Biophys. Acta 119, 649
PAUL, J. (1965). Cell and tissue culture, p. 83 (E. and S. Livingstone, Ltd., Edinburgh and London)
PEACOCK, A. C. and C. W. DINGMAN (1967). Biochemistry 6, 1818
PENSWICK, J. R. and R. W. HOLLEY (1965). Proc. Natl. Acad. Sci. (Wash.) 53, 543
PETERSON, E. A. (1970). Laboratory techniques in biochemistry and molecular biology, Vol. 2 (North-Holland, Amsterdam-London)
PETERSON, E. A. and H. A. SOBER (1962). Methods in enzymology, Vol. 5, p. 3 (Academic Press, New York and London)
RAJBHANDARY, U. L. and S. H. CHANG (1968). J. Biol. Chem. 243, 598
RAJBHANDARY, U. L., S. H. CHANG, A. STUART, R. D. FAULKNER, R. M. HOSKINSON and H. G. KHORANA (1967). Proc. Natl. Acad. Sci. (Wash.) 57, 751
RAJBHANDARY, U. L., R. D. FAULKNER and A. STUART (1968). J. Biol. Chem. 243, 575
RANDERATH, K. (1967). Arch. Biochem. 21, 480
RANDERATH, K. and E. RANDERATH (1967). Methods in enzymology, Vol. 12 Nucleic acids, Part A, p. 323 (Academic Press, New York and London)
RAZZELL, W. E. and H. G. KHORANA (1961). J. Biol. Chem. 236, 1144
REDDI, K. K. (1959) Biochim. Biophys. Acta 36, 132
RICHARDSON, C. C. (1965) Proc. Natl. Acad. Sci. (Wash.) 54, 158

Roblin, R. (1968) J. Mol. Biol. *31*, 51

Rogg, H. and M. Staehelin (1969) Biochim. Biophys. Acta *195*, 16

Rushizky, G. W. (1967) Methods in enzymology, Vol. 12, Nucleic acids (46) (Academic Press, New York and London)

Rushizky, G. W., E. M. Bartos and H. A. Sober (1964) Biochemistry *3*, 626

Rushizky, G. W. and H. A. Sober (1962) Biochim. Biophys. Acta *55*, 217

Rushizky, G. W., and H. A. Sober (1963) J. Biol. Chem. *238*, 371

Sanger, F. and G. G. Brownlee (1967) Methods in enzymology, Vol. 12 (43) (Academic Press, New York and London)

Sanger, F., G. G. Brownlee and B. G. Barrell (1965) J. Mol. Biol. *13*, 373

Sanger, F., and H. Tuppy (1951) Biochem. J. *49*, 463

Schweizer, M. P., G. Chheda and R. H. Hall (1968) Abstr. 156th Meeting of the Am. Chem. Soc. (Atlantic City, N. J., U.S.A.)

Singer, M. F. and J. K. Guss (1962), J. Biol. Chem., *237*, 182

Smith, J. D. (1967) Methods in enzymology, Vol. 12, Nucleic acids (Academic Press, New York and London)

Sober, H. A. (1968) Handbook of biochemistry (The Chemical Rubber Co., Cleveland, Ohio, U.S.A.)

Staehelin, M. (1961) Biochim. Biophys. Acta *49*, 20

Stanley, W. M. (Jr.) and R. M. Bock (1965) Biochemistry *4*, 1302

Sulkowski, E. and M. Laskowski (Sr.) (1962) J. Biol. Chem. *237*, 2620

Szekeley, M. and F. Sanger (1969) J. Mol. Biol. *43*, 607

Takanami, M. (1967) J. Mol. Biol. *23*, 135

Tomlinson, R. V. and G. M. Tener (1963) Biochemistry *2*, 697

Tyndall, R. L., K. B. Jacobson and E. Teeter (1964) Biochim. Biophys. Acta *87*, 335

Uziel, M. (1967) Methods in enzymology, Vol. 12. Part A, p. 407 (Academic Press, New York and London)

Uziel, M., C. K. Koh and W. E. Cohn (1968) Anal. Biochem. *25*, 77

Venkstern, T. B. and A. A. Baev (1965) Ultraviolet spectra of some nucleotides, nucleosides, bases and oligonucleotides (Nauka, Moscow)

Volkin, E. and L. Astrachan (1956) Virology *2*, 149

Warner, J. R., R. Soeiro, H. C. Birnboim, M. Girard and J. Darnell (1966) J. Mol. Biol. *19*, 349

Watson, J. D. and F. H. C. Crick (1953) Nature (Lond.) *171*, 737 and 964

Williams, R. B. and R. M. C. Dawson (1952) Biochem. J. *52*, 314

Williamson, R. and G. G. Brownlee (1969) FEBS Letters *3*, 306

Wyatt, G. R. (1951) Biochem. J. *48*, 584

Yaniv, M. and B. G. Barrell (1969) Nature (Lond.) *222*, 278

Yaniv, M., A. Favre and B. G. Barrell (1969) Nature (Lond.) *223*, 1331

Zachau, H. G. (1969) Angew. Chem. (Internat. Edit.) *8*, 711

Zachau, H. G., D. Dutting and H. Feldmann (1966) Hoppe-Seylers Z. Physiol. Chem. *347*, 212

Zamir, A., R. W. Holley and M. Marquisee (1965) J. Biol. Chem. *240*, 1267

Subject index

Absorbancy units 22
acid DEAE-cellulose column, *see* chromatography
acid fuchsin 66
acrylamide gels, *see* electrophoresis
alkali-resistant dinucleotides, mobilities 205–206
alkaline hydrolysis 43
alkaline phosphatase, *see* phosphatase
ammonia 198
A's, cleavage 199–200
Aspergillus 46
ATP, γ-^{32}P 226–227
autoradiography, *see* radioautography

Bacteriophage Qβ 226, 231
band width in paper electrophoresis 64
base-pairs 13, 150, 181–185, 187
base-specific cleavage 188, 196–200
BBOT 77, 122
benzoylated cellulose 203
binding forces on columns 20
Biogel 35
blocking reagents 102
blotting 73
^{3}H-borohydride 226
buffers for electrophoresis, *see* electrophoresis

*iso*butyric acid–ammonia–water 34

Capillary tubes 76
carbodiimide
  water-soluble 155–163, 196–198
    removal of 198
carrier RNA 76
Cellogel 70
cellulose acetate 68, 70
chromatography, column
  benzoylated cellulose 203
  DEAE-cellulose 25
    55 °C 30–31
    acid 28, 34
    ammonium carbonate 28
    desalting 35–38
    neutral 27–29, 32, 34

  DEAE-Sephadex, acid 27, 32, 34
  Dowex-50 36
chromatography, DEAE-paper
  pH 8.0 217
  salt-gradient 153
chromatography, ion-exchange on columns 20
chromatography, paper 34, 54
  descending 207
  two-dimensional 43

chromatography, solvents for 34, 45, 206
chromatography, thin-layer 19, 136–137
  sensitivity 43
  two-dimensional 137–140
CO_2 injection 61
complete digestion, *see under name of enzyme*
composition
  and position 85
  of oligonucleotides 76
coolant for electrophoresis 55
C's, cleavage 196–198

DEAE-paper, *see* paper
degassing 29
denaturation 187
depurination 75
derivation of sequence, *see* sequence
desalting 35–37
desiccator 76
diagonal 218–220
dialysis 37
dimethylsulphate 165
dinucleotides, resistant 48, 205–206
dyes 66, 73
DNA polymerase 13

E. coli
  ^{32}P-labelling of 241
  strains and 5S RNA 180
electric blender 137
electric shock 61
electro-endosmosis 67
electrophoresis 48, 54–66
  acrylamide gel 100, 243–247
  buffers for 64–66, 72
  cellulose acetate pH 3.5 70–73
  coolant for 54–55
  DEAE-cellulose
    pH 3.5 65, 78–79
    7.5% triethylamine carbonate 123
  flat-plate 55

marker dyes 66
papers 54–64
power supply for 58–64
room 63
safety and 58–64
tanks, 'hanging' 55–56, 59–61
  cooling-coil for 60
  electrodes for 61
tanks, 'up-and-over' 56–58, 62, 66
two-dimensional fractionation, *see* two-dimensional elution
elution
  from acrylamide gels 247
  from paper 76
  from thin layers 141
    device 141
end groups 213–220
  of 5S RNA 117–119
5'-end group 226
end products 100
endonuclease, *Neurospora crassa* 215
enzyme digestion, *see under name of enzyme*
  summary 250–255
enzyme specificity 39–40
evolution 186
exonuclease, *see* phosphodiesterase
extinction coefficient 37, 241

Film, X-ray 75
finger ribonuclease, *see* ribonuclease
fingerprinting, 16S and 23S rRNA 82–89
7% formic acid system 86

Geiger counter, portable 137
gel filtration 35

Hair dryer 64
hair-pin loops 187
half molecules 28–30, 101
high-voltage electrophoresis, *see* electrophoresis

homochromatography 130–140, 188–192
  DEAE-paper 132–136, 190
  DEAE thin layers 136–140, 190
    equilibration 138
homogenizer for thin layers 137
homomixture *a* 133
  b 140
  c 140
hypochromicity 38

Ink, radioactive, *see* radioactive ink
internal standards for yields 123
ionophoresis, *see* electrophoresis
IUPAC-IUB rules 16, 237

Liquid scintillation counting 122
loading 64–66

'Macaloid' 24
maps
  of P-RNase digests 88–89
  of T_1-RNase digests 83–87
mass spectroscopy 52
metal ions in paper 214
methyl groups in rRNA 223
N^7-methylguanylic acid 165
methyl orange 66
micrococcal nuclease, *see* nuclease
minor bases 202–212
  ala, val, tyr tRNA 44
  clustering in rRNA 214
  identified since 1965 47
  m^1A and m^6A 173
  mobilities 205
  pK_a 48
  tRNA^{Tyr} 208–210
  unstable to alkali 46
MN 300 cellulose 137
MN 300 DEAE 137
MN 300 HR cellulose 137
M values
  pH 1.9 system 90

pH 3.5 system 111, 113
myosin 228

Nearest-neighbour 232–233
neutral DEAE-cellulose columns, *see* chromatography
nomenclature 12, 16–17, 237–239
nuclease, micrococcal 41, 51, 78
nucleoside 3′,5′-diphosphates 214
  mobilities of 215

Overdigestion 80
overlapping, *see* sequence
Oxoid 70

Pancreatic ribonuclease, *see* ribonuclease
paper
  chromatography, *see* chromatography
  electrophoresis, *see* electrophoresis
  post-slip 63
  Whatman No. 1 64
    No. 3 MM 64, 73
    No. 540 64
    DE 81 (DEAE) 68, 73–75
partial
  digestion, *see under name of enzyme*
  methylation, *see* methylation
phenol 49, 227
phosphatase, alkaline 49, 51, 52, 116
  digestion on paper 218
  heat inactivation 81
phosphate, inorganic, mobility of 216
phosphodiesterase
  snake venom
    complete digestion 48, 80
    partial digestion 40, 49–51, 109–117, 189
    pppNp hydrolysis 216
  spleen, partial digestion 81, 89–99, 120–122
phosphokinase, polynucleotide 215, 225–231

phosphorylase, polynucleotide 51
pK_a of bases 21, 48
polymerase, see RNA and DNA
polynucleotide phosphokinase, see phosphokinase
polynucleotide phosphorylase, see phosphorylase
polythene sheets 76
position and composition 85
potatoe apyrase 216
propan-2-ol–ammonia–water 206
propan-2-ol–HCl–water 45, 206
n-propanol–ammonia–water 34
pyrophosphate, mobility of 216

Radioactive ink 75
radioautography 68, 75
rare bases, see minor bases
recrystallisation
  of acrylamide 245
  of methylene bisacrylamide 245
replicase, bacteriophage Qβ 231
resolution on paper electrophoresis 64
ribonuclease
  CB 47
  finger 37
  pancreatic
complete digestion of RNA 22, 69, 118–120, 196–197
  complete digestion of T_1-oligonucleotides 48, 77–78
  partial digestion of RNA 24, 150–155
  partial digestion of oligonucleotides 201
  purification 23
spleen acid 155, 200–201
T_1
  complete digestion of P-RNase oligonucleotides 48, 78
  complete digestion of RNA 21, 69, 83, 103–106

'half' molecules 28, 148–150
partial digestion 23, 30, 82, 143–148
purification 220
and phosphatase, complete digestion 22, 70
T_2 46–48, 207, 210–211
U_2 199–200
ribosomal RNA, see RNA, ribosomal
RNA
  bacteriophage
    R17, f2 190–194, 248–249
    MS2 122
  E. coli
    ^{14}C-methyl labelled 249
    ^{3}H-methyl labelled 249
    ^{35}S labelled 250
  labelled in vitro 225–234
  mammalian, ^{32}P-labelled 248
  mononucleosides in 11
  polymerase 14
  purity for sequencing 129
  ribosomal 5S 100–186
    gene duplication 185–186
    of humans 186, 229
    of rabbits 228
    purification 243–247
    secondary structure 181–185
    sequence homologies 185–186
    sequence variations 180–181
  ribosomal 16S and 23S, methyl groups 223
    methylated sequences 220–224
    P-RNase digests 86–89
    T_1-RNase digests 82–86
  6S 102, 194–196, 216
    purification 243–247
  transfer
    and labelled minor bases 202–212
    ^{32}P-labelled, purification 203
    ^{32}P-labelled, sequenced 204
    ^{32}P-labelled, tyrosine suppressor 208–210

unlabelled 18–53
yeast, ^{32}P-labelled 247

Sample loading on paper 64–66
secondary
  splits 107–109
  structure of 5S RNA 181–185
sensitivity of detection 42
Sephadex 35
sequence
  derivation of 101, 173–180, 192–196, 232
  methods, summary 250–255
  of large T_1 end-products 187–201
  of oligonucleotides, summary 252
  overlapping of 101, 130
  solutions for, summary 254–255
  variations in 5S RNA 180–181
snake venom phosphodiesterase, *see* phosphodiesterase
solvents, *see* chromatography
specificity, *see* enzyme
spectrum, *see* ultraviolet absorption
spleen
  acid RNase, *see* ribonuclease
  phosphodiesterase, *see* phosphodiesterase
synchrony 231

Tanks for electrophoresis, *see* electrophoresis
Takadiastase powder A 47
template, thin-layer 137
thin layers, *see* chromatography and homochromatography
4-thiouridine 209
  and light 203

cross-bridges 203, 211–212
T_1-ribonuclease, *see* ribonuclease
triethylamine carbonate 36, 76, 109, 123, 142
tRNA, *see* RNA, transfer
Tween-80 81
two-dimensional fractionation
  homochromatography 130–140, 188–192
  ionophoretic 67–99
  paper 43
  summary 251–252

Ultraviolet absorption 18
  spectrum 38, 50, 52
urea buffer 70
U_2-RNase, *see* ribonuclease
Ustilago 49, 199

Varsol 55
venom phosphodiesterase, *see* phosphodiesterase

Water-soluble carbodiimide, *see* carbodiimide
wet-loading 66
Whatman, *see* paper
white-spirit 58, 61, 63

X-ray film, *see* film
xylene cyanol FF 66

Yield
  estimation of 77
  molar 38
  relative molar 37–38, 122–130

Randall Library – UNCW
QD434 .B76 1972
Brownlee / Determination of sequences in RNA
NXWW

304900191740